"ABOUT"

THE POWARS II PALEOINDIAN RED OCHRE MINE IN EASTERN WYOMING

Cris - Keep Lookin

Compiled for
George C. Frison

George M. Zeimens

For a short video, see:
https://vimeo.com/988958282/6deb99a0f9

Printed by
Hynek Printing, LLC
Richland Center, Wisconsin

ISBN: 978-0-9743178-7-8

Contents

Preface

George M. Zeimens

Archeology is not an exact science. Interpretations of human behavior from what is found at a site rely heavily on comparisons of similar findings from other sites. Based on a body of accumulated data from numerous sites, patterns of behavior begin to emerge, at least in the eye of the beholder. The problem is, human behavior does not always conform to a pattern. Human beings often do irrational things that cannot be explained by material items left at a scene. Most often data used to explain human behavior at a site is based on nonperishable remains. The further a site extends back in time, the fewer perishable artifacts are available. At the oldest sites, such as Powars II, nonperishable data consist primarily of stone and bone. Much can be learned about past activities from stones and bones, but lacking is the ability to extract from them that part of humanity that perpetrated the acts represented by the artifacts, i.e. the thoughts, emotions, meaning and purposes that constituted the lives of the site occupants and their way of viewing the world.

At just about every site that I worked on with George Frison, as we were about to complete our investigations either at the site or during analyses in the lab, he would shake his head and solemnly mutter, "We are missing something.". He made that remark on numerous occasions as we made one astonishing discovery after another while we were exploring the archeological deposits at this site, the Powars II Paleoindian Red Ochre Mine. Almost immediately, it became clear that we were dealing with a discovery the likes of which we had never seen before, nor had we ever imagined that such a bonanza of archeology existed.

Without George Frison's involvement, this site would likely have been destroyed, which would have been a tremendous loss to archeology, as well as to numerous areas of humanity. His friends will tell you that after his wife, June, passed in 2011, although he continued on with research, he became rather despondent and his health began to fail. When after languishing for twenty-five years the red ochre mine finally became available for archeological study, he responded enthusiastically. But it wasn't until after digging into the eroded talus deposits below the ochre mine, and learning that the site was the most exciting Paleoindian treasure trove that he had ever seen, that he perked up and became his old, dynamic self again. His friends will tell you that the site undoubtedly extended his life by several years. He had just started writing this book, and while on his death bed it was his request that we bring it to completion. So we have, but it is has been difficult for us to present the material with the high standard of writing that he was known for in the many books that he authored during his career. Deficiencies and errors in writing and content that readers may find herein are not his, but are ours.

It won't take the reader long to realize that

Powars II is one of the most exciting Paleoindian sites in North America. Most of the site remains to be explored and what, and just how much it has to offer to the archeological record, remains to be determined. One of the reasons that George Frison thought that this kind of book would be important and appropriate, is that it provides an opportunity for amateur archeologists, collectors, educators and other interested parties that do not subscribe to professional journals to learn about this amazing site. After all, the site constitutes a part of everyone's heritage, not just the professional community.

George M. Zeimens

Chapter 1

Introduction

George M. Zeimens

Powars II is a Paleoindian red ochre mine. It is the only Paleoindian red ochre mine that we know of in North America. The site is a record of over 3,000 years of Paleoindian occupation and activity in this area. All of the Paleoindian cultural complexes known on the Western High Plains are represented in the site.

Because the site is of international interest, George Frison felt that it was necessary to prepare a book "about" the history and development of the Powars II site. It is about what we know about the site at this point in time, and its potential for future research. It is a story about how the site was discovered and rescued while literally on the brink of destruction, and about how fortuitous circumstances have played a role in preserving and securing the site. This book is not a site report. Research at the site is in initial stages, and we have gleaned but a small amount of technical data such as unit maps, unit profiles and other data pertinent to the formation and archeology of the site. Exceptions are a thorough analysis presented in chapters 7. and 8. of a projectile point type from the site that we initially referred to as Clovis. Those studies were necessary in order to clarify the cultural context of the site; in particular whether or not it is a Clovis site. Also, in Chapter 6, a detailed analysis of red ochre, and of a newly discovered type of toolstone located near the ochre mine is presented, which adds credence to the significance of the site.

Up to the present time, most of our work has been to secure the site, clean up the surrounding area, and control site erosion while attempting to learn the spatial and temporal limits of the deposits. In doing so, we identified numerous logistical problems that needed to be addressed. The site is located on a relatively steep hillside on the edge of historic industrial copper and iron mines. Historic copper and iron mining activities took place above, below and on both sides of the ochre mine. Without a doubt, some unknown portion of the ochre mine was damaged or even destroyed by historic mining. The hillside was littered with mine spoils and other evidence for historic activity including old prospecting adits. A shallow trench that extends through the site area appeared to be a historic adit, or drift. The area at the base of the hill below the ochre mine contained tons of eroded site deposits mixed with historic mine tailings and railroad debris. It was obvious that erosion of the site was a problem, and that the historic debris located on and above the site needed to be cleaned up before investigations could proceed.

In removing the talus at the base of the hillside that contained eroded deposits from the ochre mine, problems arose concerning the cultural context of the site. We recovered numerous Paleoindian projectile points and none that were more recent. One of the projectile points appeared to be of the Clovis type, but not quite. Also, the talus contained numerous utility types of tools that indicated that mining was not the only activity that had taken place by prehistoric occupants. Cleaning up the talus and controlling erosion turned out to be monumental tasks that required monumental efforts presented in later chapters in this volume.

Because we had learned the degree of uniqueness and rarity of the find, George Frison directed that we prepare a paper for the professional community. Unfortunately, and as Frison readily admitted, some of our observations and interpretations of certain aspects of the site presented in our article in *American Antiquity* were premature (Frison et al. 2018). For example, we initially identified the majority of projectile points from the site as Clovis, and that they are associated with classic Clovis blade core technology. As we learned later, that was not the case, and that there is but very little evidence for a Clovis presence at the site. That article, along with a later article in the *PNAS Journal* (Pelton et al. 2022), and earlier references to the site (Kornfeld et al. 2010 pp 461, Frison 2004 pp 219, Tankersley et al. 1995, Stafford et al. 2003), resulted in the site becoming known as a major Clovis site. Information presented in this book serves to correct the Clovis site misnomer, and other flaws in those articles.

As we worked at the site, we made one amazing discovery after another. As we searched for a campsite on the terrace below the ochre mine, we discovered a toolstone quarry that consisted of a unique material unfamiliar to us. That quarry is related to the ochre quarry, and will become an important factor in future lithic research considerations at sites in the area. Another important discovery in a deep test in the terrace, is numerous artifacts with dates that are older than Clovis. So far the early dates are not directly on the artifacts, but are on the sediments that contain the artifacts. However, they do date an old component of the site that may prove to be cultural. We hope to acquire more reliable dates as we continue our work in that area.

Chapter Two presents the ecological and cultural context of the site within an area of hills that constitute a rich complex of numerous interrelated archeological sites. That, and the context of the two minerals, red ochre and the discovery of the new type of toolstone, are important elements in the story of the site. Chapter Six presents the geology of the area and serves to define the context of the ochre and toolstone.

Investigations at Powars II are just beginning, and this book serves as a prelude to information from future, more in-depth research into the many aspects of this astonishing site.

Chapter 2

Natural and Cultural Contexts of the Site

George C. Frison,
George M. Zeimens,
and Mark Russ

Powars II does not stand alone, but is a part of a number of previous archeological investigations in the region that have contributed to a growing body of knowledge pertaining to Paleoindian life ways. The Powars II red ochre mine situates geographically almost in the center of several notable Paleoindian sites including Anzick (Lahren and Bonnichsen 1974), Mill Iron (Frison 1996), Agate Basin (Frison and Stanford 1982), Carter-Kerr McGee (Frison 1984), Hanson (Frison and Bradley 1980), Horner (Frison and Todd 1987), Casper (Frison 1974), James Allen (Mulloy 1959), Hudson-Meng (Agenbroad 1978), Fetterman also known as LaPrele (Mackie et al. 2020), Hell Gap (Larson et al. 2009), Scottsbluff Bison Quarry (Barbour and Schultz 1932), Jones-Miller (Stanford 1974), Lindenmeier (Wilmsen and Roberts 1978), Dent (Figgins 1933) and many others (Figure 2-1.). All of these sites and others are located on the plains east of the Continental

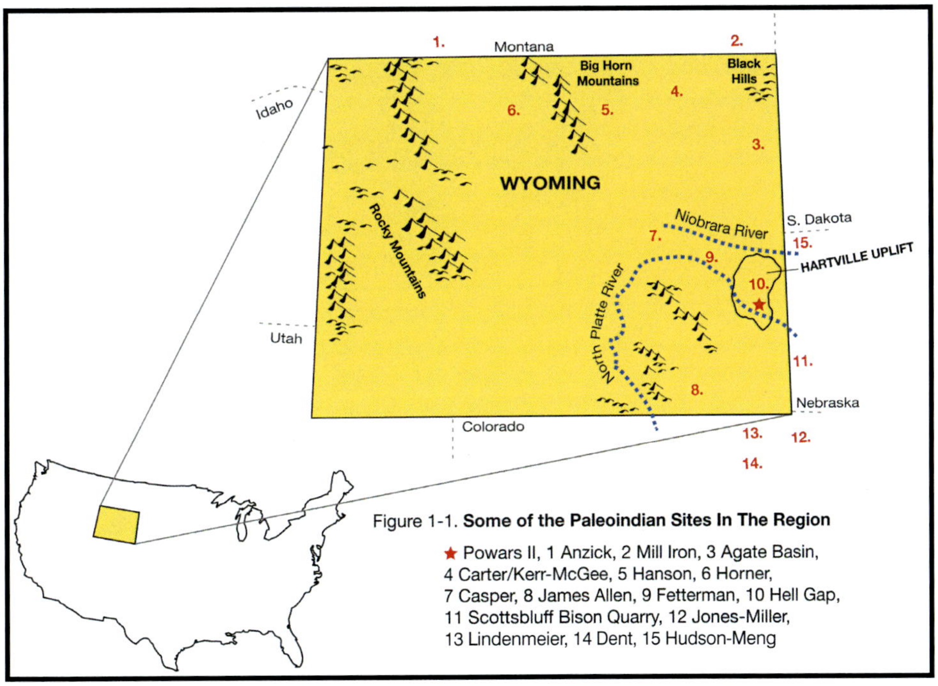

Figure 2-1. Location of Powars II in relation to some other major Paleoindian sites in the region.

Divide and the Rocky Mountains. Red ochre was found to be present in many of them, suggesting that red ochre was a substance that played a role in the everyday lives of the earliest people to occupy this region.

The ochre mine is located in the abandoned Colorado Fuel and Iron (CF&I) iron ore mine in the southern portion of a range of hills known as the Hartville Uplift. The Hartville Uplift is a low range of rocky and rather sparsely tree-covered hills. The Uplift is roughly forty-five miles long and twenty-five miles wide situated on a northeast/southwest tangent. It connects with a larger and higher range of hills, the Laramie Range to the west, and it is only an hour drive northeast to the Black Hills. The Uplift is bordered by the North Platte River on the west and south, the Niobrara River on the north, and by mainly open plains and sand hill country to the east. At a relatively moderate elevation of roughly 4,500 to 5,500 ft. (1,600m to 2,400m), and with numerous trees and springs at the fringe of the open plains, the entire uplift provides the same natural resources as a foothills ecological zone (Figure 2-2.).

This range of hills abounds in archeological sites that date from the Paleoindian era through Historic times. Much of the archeology of the area remains unexplored and undocumented mainly due to the fact that the majority of the land is privately owned. Most of the area lacks oil, coal and gas, so consequently few cultural resource management studies(CRM) have been conducted here compared to other regions. One exception is the National Guard range which has an active CRM program.

Europeans did not populate the area until relativity late, so its history only extends back 180 years or so. Historic trails through the hills include the Trapper's Trail, the Immigrant Trail, the Cheyenne/Deadwood Stage Trail and the Texas Cattle Trail. The area contains numerous pioneer-era mines including the Silver Cliff silver mine on the north end near the town of Lusk, a mica mine in McCann Pass, copper mines in the southern portion of the hills near the town of Guernsey and in the central area of Muskrat Canyon, and the large Colorado Fuel and Iron (CF&I) Sunrise iron mine near Hartville. Many areas, especially in the southern portion of the uplift, are littered

Figure 2-2. Typical environment in the Hartville Uplift range of hills.

with old prospect adits. All of those mines are now defunct. Military installations include the historic Government Farm (an old Fort Laramie outpost), and in the 1930's a major Army National Guard facility was established and still functions in the southwestern area of the hills. The first cattle herd arrived in the hills north of the North Platte river in 1874 (Bronson 1962). Raising cattle remains the main industry in the area today, and the Uplift is sparsely populated with small ranches.

The Uplift is most renowned for prehistoric sources for high quality toolstone, mostly for what is popularly known as Hartville Uplift chert found in extensive outcrops of Mississippian and Pennsylvanian limestone. This fairly distinctive stone varies from shades of brown, yellow and red and often, but not always, includes dendrites. Several varieties of chert are also present in other geological formations in the Uplift, and much of it is not always as distinguishable as Hartville Uplift material as some may claim.

A well-known toolstone site in the Uplift is the Spanish Diggings quartzite quarry (Holmes 1919, Saul 1969, Reher 1991). The site name is a misnomer, a result of early white settlers, after first seeing the huge pits and piles of tailings, surmised that they must represent Spanish miners digging for gold or silver (Gilder 1908, 1909). Of course, Spanish explorers never traveled this far north. Spanish Diggings is a source of quartzite (actually a silicified sandstone) of the Cloverly-Morrison formation, and the site consists of numerous deep pits and crevices surrounded by several acres (hectares) of tipi rings. A rock alignment in the form of a human effigy is located in the area of the rings. The diggings are deep and extensive, and represent an impressive amount of work by prehistoric miners (Figure 2-3.). Spanish Diggings does not stand alone, as there are many sources of the same silicified sandstone in other areas of the uplift including other geological formations. Most of the other sources also contain evidence of quarrying activity. In all, there are literally hundreds, if not thousands, of toolstone quarries in the Hartville Uplift with numerous varieties of chert and quartzite. A great many of them are quite extensive, formal sites that exhibit evidence of intensive mining, while some are more informal

Figure 2-3. Prehistoric pits and spoils piles at the Spanish Diggings quarry.

locations where materials were merely collected from the surface. Secondary sources are found in major drainages, and on Pleistocene terraces that extend a considerable distance to the east onto the plains of Nebraska, and some of those have been quarried prehistorically (Miller 2010, Zeimens 1999). Also, a mixture of Hartville Uplift material, both chert and quartzite, that originated in primary context in various geological formations in the hills, is found in secondary deposits in the Arikaree formation, and many of them are also primitive quarries.

Materials from the many toolstone sources from various geological contexts, both in the Hartville Uplift and from other regions of the Western High Plains, overlap in color and texture visually. Consequently, working with chipped stone material from archeological sites can be very tenuous, as was the case in identifying the lithic artifacts at the Powars II site. Because the quarries are so numerous, to attempt to distinguish between them by chemical and/or mineral analyses would require a monumental and extremely expensive project. Samples would not only be required from each source, but numerous samples from within each source would also be necessary (Christensen 1990, Wray 1988). A new type of toolstone discovered in the terrace below the Powars II Paleoindian red ochre mine, further complicates efforts to trace the source of lithic artifacts found at archeological sites (see Chapter 3.).

As might be expected in such an edge-area environment, there are many campsites in the Uplift, most of which are located in well sheltered areas along streams and near springs. Evidence for Paleoindian occupation exists at numerous camp sites. Hell Gap, located roughly 5km (miles) north of Powars II, is a deeply stratified campsite with numerous components that, as with Powars II, represent a variety of Paleoindian complexes. Discovered in the 1950s, Hell Gap is the premier Paleoindian site on the Western High Plains, as it was the first site where multiple cultural components were found in stratified, datable deposits (Irwin et al. 1973, Agogino 2009). The Paleoindian levels are superimposed by multiple Archaic and Late Prehistoric components, including a Woodland period ceramic level (Zeimens et al. 1995). Research is ongoing at Hell Gap (Larson et al. 2009).

Another camp that has the potential to become a major Paleoindian site, is the Hageman Narrows site located roughly 4km due east of Powars II. Like Hell Gap, Hageman Narrows has numerous Late Prehistoric and Archaic components situated above several Paleoindian levels; there are several levels stratigraphically beneath two levels dated at over 10,000 rcybp. The lower levels are within the water table and remain to be thoroughly investigated (Zeimens and Alkire 2005).

Mammoth remains have been found at four locations in the Uplift. A large male was salvaged between flash flood events at the Jewett site near the abandoned railroad town of Wendover on the southwestern edge of the uplift. Two stone scraping tools located in the bone bed appeared to be associated with the bones, but it is possible that they predate the level. The bones were mostly in poor condition and badly broken, so it was not clear whether they represented a kill, or perhaps a scavenged carcass. One horse metatarsal and several camel cervicals were also present in the same level as the mammoth bones. There are very pronounced cut marks on one of the camel cervicals (Zeimens 2001). Vance Haynes provided a radiocarbon date for the bone bed at 11,130 ± 100 BP (AA-21086), and we obtained a date of 11,430 ± 50 BP (Beta 107343) (Figure 2-4.). Another mammoth site located at the base of Rawhide Butte was investigated by Henry and Cynthia Irwin, but it was not determined if the find represented cultural activity (Dennis Stanford, personal communication 1986, Agogino 2009 pp 318-319). Yet another mammoth find investigated by the Irwins located at Box Elder Springs apparently did not include cultural activity (Haynes 1967). Mammoth bones still occasionally erode from that site and more investigations of those deposits are warranted. Following a recent flash flood, the proximal end of a mammoth humerus was found in the bottom of Cottonwood Draw by the landowner, but its origin was not determined. Cottonwood Draw drains the Hell Gap area, and in the 1940s two Clovis projectile points were found by local ranchers, Jim Finn and his sister, near Woodworth Spring. We recently conducted a test excavation at that location, and found some very fragmentary bones that we tentatively identified as mammoth. None of the mammoth bones from any of these sites show

Figure 2-4. Looking down at excavations at the Jewett Mammoth site.

any degree of mineralization, and all of them appear to be of Late Pleistocene/Early Holocene age.

Caves and rock shelters are also plentiful in the Hartville Uplift hills. Salvage excavations were conducted at two sites, Bass-Anderson Cave that rested in the path of destruction by limestone mining, and the Sommers site which was in imminent danger from looters. Bass-Anderson contained stratified, Early Archaic deposits along with Late Paleoindian components representative of Hell Gap, Cody and Lovell Constricted complexes. The Sommers shelter contained Late Prehistoric components including a Woodland ceramic level, Late Archaic components, and at least one Paleoindian level that yielded a Pryor Stemmed projectile point. The bottom most deposits at Sommers were not reached due to heavy rockfall below the Pryor Stemmed level (Zeimens 2002 and 1996). There is a unique red ochre pictograph on the wall of Sommers shelter that depicts a grizzly bear and two human stick figures (Figure 2-5.). The only other example of prehistoric rock art in the Uplift that we are aware of consists of a large handprint painted in red ochre. That pictograph is located in a small, prehistoric mine adit or hole, about 1m in diameter, placed in the side of a vertical limestone exposure. The hole is actually a small chert quarry where a vein of chert had been extracted for a distance of 2m into the rock wall. The handprint is on the roof of the hole (Figure 2-6.).

There are also many hundreds of stone circles or tipi rings in this range of hills, most of which apparently represent camps of various lengths of duration. They are present in various landscapes from ridge tops to the valley floors. None of the circular stone features appear to be older than Archaic times, although very few have been systematically investigated and dated. Ceramics are represented in several of the sites, but none have been demonstrated to be in direct association with the circles. There are stone circles at the Hell Gap site, but they are definitely not related to the Paleoindian complexes (Zeimens et al. 1995, Haynes 1965). A unique group of stone circles located on the hill above the Powars II site will be discussed later in Chapter 11, this volume.

Many of the open air camps contain culturally stratified deposits, and some were clearly occupied

Figure 2-5. A pictograph in red ochre at the Sommers site.

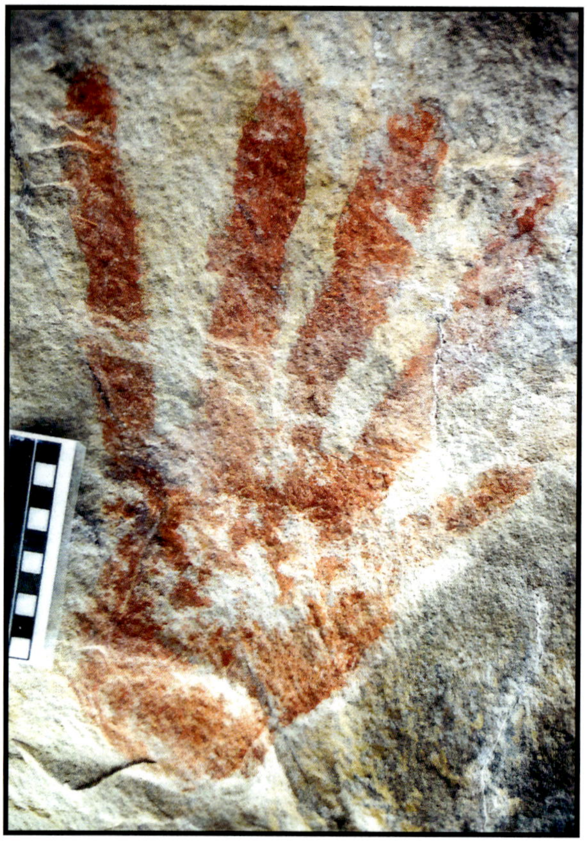

Figure 2-6. Large hand print in red ochre on the roof of prehistoric mine adit.

repeatedly by the same group on an annual basis. The stratified Late Archaic, Late Prehistoric, and Protohistoric Fisher site located in Cottonwood Draw, contains evidence for contact between Central Plains and Western Plains people, where Shoshone and Dismal River types of ceramics were found together in a single pile next to a fire hearth. The Meyers site is another important Late Period and Archaic site with a neatly stratified series of fire pits (Figure 2-7.). The lowermost pit was associated with two corner tang knives along with other Late Archaic period artifacts. Woodland Period pottery has also been found in several camps in and near the hills (Zeimens and Alkire 2005).

Several early pioneer or settler grave sites in the area include the site of the notorious outlaw Mother Feather Legs, the Brownrigg Cemetery, Box Elder Springs, Immigrant Hill, Wendover and Hartville cemeteries, and the Hageman home and dugout and infant graves. The Late Prehistoric remains of a teenage girl was found in a road cut on the southeastern fringe of the hills. Two projectile points located in the area of her abdomen indicated that she had been intentionally killed.

Figure 2-7. A neatly stratified series of fire pits at the Meyers site.

The Wyoming State Historic Preservation Office presented her remains to the tribes at the Wind River Indian reservation before they could be fully analyzed (Zeimens nd).

The Hartville Uplift afforded shelter, water and a variety of food and other resources important to prehistoric people that dwelled in the hills. But there can be no doubt that the main attraction to the area was the abundance of, and easily available, toolstone and red ochre. Red Ochre has been found to be present in many of the prehistoric sites in the area, mostly in the form of red stains in the soil, or small hematite pebbles.

Powars II contains the only known evidence for how ochre was mined and processed during Paleoindian times. The context of the artifacts suggest that ritual/spiritual activity, was a significant part of the cultural milieu of the site. Projectile points from the site represent virtually all of the known cultural complexes in the High Plains area including Goshen, Midland, Plainview, Folsom, Agate Basin, Hell Gap, Alberta, Frontier and possibly Clovis. The collection also includes an aberrant projectile point type that we are calling the Sunrise type (see Chapters 7. and 8. this volume). Our investigations at the site at this point in time are in initial stages, but we have learned that the site has the potential to address chronological problems and other issues with regard to Early Paleoindian complexes. The deposits in the ochre mine feature, a narrow linear trench that extends into a hillside (see Chapter 5.), are mixed from later groups digging through deposits left by earlier miners. Consequently, so far no absolute dates in direct association with the various projectile point types and other artifacts have been obtained. One of the most intriguing aspects of the site, are dates associated with cultural remains from deeply buried deposits in the terrace below the ochre mine that are older than Clovis.

As explained above, the ochre mine is part of a large complex of Paleoindian sites in the hills that are interrelated in that, they represent the same complexes but at different types of sites at various locations. Most of the Powars II lithic artifacts are made from toolstone found in the Hartville Uplift. A large percentage of the toolstone that the Powars II artifacts are made of can be traced to a newly discovered source located in the valley just below the ochre mine, as well as from nearby sources in the outlying hills. The material in the new source is of a different type and geological age of any material previously identified in the Uplift, or elsewhere on the Western High Plains (see chapter 3.). A comparison of material from this toolstone source with materials from other sites in the region, has the potential to enhance areas of research such as long-distance travel, trade and other aspects of Paleoindian cultures.

Chapter 3

Discovery and Initial Investigations

George C. Frison and
George M. Zeimens

Mr. Wayne Powars and his father were both artifact collectors. Several years ago they located a Folsom site near Kersey, Colorado, and collected and recorded that site with the help of professional archeologist Dr. Frank Roberts, Chief Archeologist at the Smithsonian. That site was named the Powars site. Subsequently, we named the ochre mine the Powars II site. Powars also worked for Roberts at the Lindenmeier Folsom site in Colorado from 1935-1937 (Wilmsen and Roberts 1978 pp 9-11). He later was hired as a coach at the Sunrise Mine school located at the Colorado Fuel and Iron (CF&I) company town of Sunrise. He was drafted into the army in 1941, and served at the Pentagon in Washington, D.C. Following his discharge after World War II, he apparently continued to live in the Washington area.

The story about the Powars II Paleoindian red ochre mine actually began in the late 1930s and early 1940s when Wayne Powars, while still working as a coach at the Sunrise Mine School, discovered the site and began digging artifacts out of the edge of the deposits. We don't understand why he did not notify Roberts about the site, as he had the background to understand the significance of the find. We can only speculate that, perhaps reporting the site would have jeopardized his relationship with CF&I, owners of the mine and the school. Also, the site was located on the hillside on the north side of the railroad tracks that extended along the north side of the mine. CF&I had a safety rule that no one was to cross to the north side of the tracks. If anyone knew that he had violated that rule, his job as teacher/coach at the school may have been in jeopardy. Yet another mystery is how he was able to keep the site a secret. The ochre mine is located directly across from, and in full view of the YMCA building in the center of the mine's housing area (Figure 3-1.). There were numerous artifact collectors living in the area in the 30s and 40s, and if they had found out about the site it likely would have been exploited into oblivion. A railroad had been constructed at the base of the hillside below the site. Miners working in that area surely would have seen some of the artifacts protruding from the eroded deposits from the ochre mine above (Figure 3-2.). We can find no one alive today that lived in Sunrise or worked in the iron mine that has any knowledge about the site, or of artifacts being found in the area of the site. There was a medical doctor that lived in the nearby town of Hartville from 1900 to 1902. He had a small collection of artifacts that are strikingly similar to the Sunrise projectile points and other tools from Powars II, and they were coated heavily with red ochre. His family claimed that he had found them near Rawlins, Wyoming in 1902. The mysterious story of that collection known as the Riser Cache (Boles et al. 2008), will be presented in the following chapter.

Figure 3-1. View of the Powars II site across the roof of the historic YMCA building in the Town of Sunrise.

Figure 3-2. Ochre mine perched on the hillside, Frison below the site in 2014 contemplating initial investigation tactics.

In 1981 Frison and Dennis Stanford were working at the Smithsonian putting the finishing touches on the Agate Basin site book. One day Wayne Powars walked into Dennis's office carrying a tray of Paleoindian artifacts heavily coated with red ochre. He said that he had found them in Wyoming, and when asked from where in Wyoming, he said that the collection was from a site along the railroad between the towns of Sunrise and Hartville. He also stated that there was nothing left of the site, as it had been totally destroyed by construction of the railroad. Sometime later, Frison and Stanford traveled to the area and walked the rail line between Sunrise and Hartville, about 1.1km, and found no evidence for a site. The only place they saw that looked like it might be a good candidate for a site was in the area of the old Sunrise School. The school no longer exists, but it was built on a relatively broad terrace that extends a short distance back into the hills on the south side of the road. That terrace has been leveled by filling it with mine tailings, and it is now covered with historic debris. As we learned later, the site was located farther to the east, actually within the town of Sunrise, and close to mining equipment that was in full operation at the time that Frison and Stanford searched along the old rail line, which explains why they did not search in that area.

After walking the rail line with negative results, Frison and Stanford harbored doubts about Powars' claims, but in the summer of 1986, Powars returned to the area to attend a Sunrise School reunion. That weekend he went to the site, which was actually located within the boundaries of the iron mine, with intentions of doing some digging. The mine had ceased operation several years earlier. Upon arrival he saw that site was in imminent danger of being totally destroyed by Abandoned Mine Land Reclamation (AML), under the direction of the Wyoming State Department of Environmental Quality (DEQ). Heavy machinery used to reclaim the hillside rested right at the edge of the location where he had previously recovered artifacts. Work had been halted for the weekend, but was going to continue on Monday morning and would have extended directly through the site deposits. Powars immediately went to a phone and notified Dennis Stanford of the situation. Stanford then phoned Frison who was able to work with DEQ to halt work in that area, and to set the area aside from future reclamation plans. Reclamation had already damaged a portion of the site by dumping large rocks, mine spoils and other debris onto the site from the edge of a mine haul road that extended across the hillside above the site area (Figures 3-3., 3-4.). The amazing part of this story is that we had come within one day of losing a rare and highly significant Paleoindian site, one that had survived for over 13,000 years. Had Stanford not been in his office by the phone on that weekend, and Frison not in his office by a phone on that weekend, there is no doubt that the site would have been completely obliterated when reclamation commenced on Monday morning. Powars' quick action saved the site, but it is very unfortunate that he had not reported the site earlier.

Before he left the area, Powars pointed out to Frison the precise area of the deposits on the hillside where he had collected his artifacts. Neither of us had contact with him again before he died, so we were not able to garner more information from him about how he went about collecting from the ochre mine, or how he was able to secret the site from everyone while working in full view of Sunrise miners and residents.

Because federal AML funds were used for reclaiming the iron mine, a private consultant was hired to conduct a cultural resource study of the proposed areas of disturbance before work commenced (Whissen 1985). The ochre mine was not identified during the study. Consequently, some of the site was damaged and came close to being destroyed by the reclamation work. More importantly, had we known about the site at an earlier date, it might have been possible to secure it properly for future research. However, during this time the property fell into private hands as a result of legal issues between CF&I and the State of Wyoming. With the iron mine then being in private ownership, the site became off limits for research purposes for the next twenty-five years. During that time a large and significant portion of the intact site deposits were allowed to erode down the hillside. The only positive outcome of the situation was that the private land owner did not allow anyone else to encroach upon the site.

In order to evaluate the site, Frison had been allowed to conduct a small test excavation into

Figure 3-3. Heavy machinery about to encroach upon the site (1986).

Figure 3-4. Mine spoils intended to be pushed over the top of the site by mine reclamation.

the site in 1986 (Stafford 2003). He placed a test the length of 2m into the southeastern edge of the deposits. Today, after languishing for twenty-five years, all but about 20cm of the end the test remains revealing that since 1986, almost 2m of the eastern edge of the site deposits eroded and slid down the hillside.

Ownership of the mine changed in 2011. When We approached the new owner, Mr. John Voight, about obtaining access to the site, he was somewhat incredulous about our request to encroach upon his property. He did not know about the site because the previous owner had not told him about it. He also did not know us or anything about archeology. John did agree to let us take a look at the site (neither of us had seen the site for several years), and let him know what we wanted to do. When we said that what we wanted was to screen the talus deposits below the site, he was agreeable to that plan. After we began screening, and he saw the amount and kinds of artifacts that we were recovering, he got excited and wanted to learn more about the site, and about Paleoindians. He has since become a real archeology enthusiast, and has supported our research in numerous ways for the last several years. John is also very active in efforts to preserve both the historic and prehistoric components of the site for future generations of "researchers and humanity", (John's own words).

Investigations in the Talus

The exploratory excavation conducted by Frison in 1986, established the fact that Clovis-age artifacts exist in the deposits on the hillside some 12m to 15m above the valley floor. We knew very little about the nature of the site, but were fairly confident that it served as a Paleoindian red ochre mine. Up to 2m of prehistoric, intact deposits were exposed on the southern and eastern edges of the site on the hillside. There was a large amount of talus on the valley floor immediately below the site, over 2m deep in some areas, and for a length of 40m along the base of the hill (Figure 3-5.). The talus deposit was the result of erosion that took place during the last one hundred years, after the valley was filled to the current level to accommodate historic mine facilities. It is not known how much of the site eroded prehistorically, and was buried during leveling and filling of the area below the site by CF&I. An abandoned railroad bed extends along

Figure 3-5. The large amount of talus that had accumulated below the ochre mine.

Figure 3-6. A 1986 photograph of mine reclamation work surrounding the site, The ochre mine trench feature is clearly visible next to exposed cultural deposits on the hillside.

the base of the hill, and at one time spoils from the historic iron mine were used as ballast to build up the rail bed. During construction the prehistoric talus was apparently shoved around and pushed up against the hill on numerous occasions, causing those materials to become mixed with historic mine spoils that contained railroad and mine trash. Also, an old mine haul road extends across the hillside above the site and boulders, tailings and trash were pushed over the edge of the road down onto the hillside, resulting in additional contamination of the talus and scree deposits (Figure 3-6.).

While the landowner from 1986 to 2011 would not permit excavations, he did allow us to visit the site on several occasions, provided that we did not collect anything. On those visits we observed flakes and an occasional stone tool protruding from the talus or lying on the scree. With acquisition of the property by John Voight, and financial support from a consortium of very generous artifact collectors, we finally had an opportunity and resources to develop plans to investigate the site. We knew that we had to proceed carefully since Powars II is an extremely rare, one-of-a-kind site, the likes of which has never been excavated before. Therefore, we had no previous research model from which to work. We also knew that the talus contained artifacts, and calculated that any work in the intact deposits on the hillside above, would likely result in loss or damage to artifacts in the talus below. We concluded that our initial effort should be to screen a sample of the exposed and vulnerable talus. We hoped that in doing so we would recover a few diagnostic artifacts, and potentially ascertain a better understanding of the nature of the intact deposits above.

At that time we worked under the auspices of the Western Plains Historic Preservation Association (WPHPA), a private, nonprofit organization. That organization sponsored an educational summer youth program called Expanding Environments (EE), where high school students participated in historic preservation projects and archeological salvage excavations. We did know that the artifacts in the talus originated directly from the hillside above, but also recognized that they were contained in mixed and secondary prehistoric deposits that included historic mine tailings and trash. Consequently, little could be learned by piece-plotting every item. Screening the talus without the need to plot artifacts would be a good job for a crew like the EE high school students.

We began screening early in the summer of 2014. We started work with what turned out to be a very naive calculation that we would be finished in a few days or no more than a few weeks, and hoped to recover a few informative artifacts. As it turned out, digging into the talus was much more difficult than we had anticipated. The talus contained a mixture of historic mine spoils, mine related trash, detrital iron ore debris from the hillside, powdery iron dust or "paint" ore and chunky and slivery bits of schist, but almost no earthy soil. The deposit had become tightly consolidated almost to the same consistency as concrete, and simply could not be penetrated effectively with a trowel or even a shovel. What did work was to break it into chunks with picks. Once loosened, the chunks would crumble almost spontaneously, or with a light blow from a shovel or pick. Then the larger rocks could be sorted out and the rest could be shoveled into screens. Breaking the talus apart with picks proved to be a safe way of recovering the artifacts, as we eventually screened an enormous amount of material with only slight damage to two artifact specimens.

Screening turned out to be an equally assiduous and time consuming operation. A thick coat of pasty, red dust or a hard, metallic crust adhered to every object in the screen including small nodules of iron, slivers of schist, trash and the artifacts. The short, flat, lenticular slivers of schist were a particular nuisance in that they often took on the appearance of a projectile point or other artifact, and each one had to be closely scrutinized. Screeners had to be careful not to discard an artifact that might be mistaken for a piece of schist until the red coating was rubbed off. Some of the artifacts were so heavily encased in soft, clay-like ochre, that it appeared that they were intentionally painted and encased in ochre before they were left in the mine. We had noticed the extreme amount of ochre coating on the artifacts at the time of Frison's 1986 test excavation (Tankersley et al. 1995). The screening process was slow, in that we could only screen one shovelful at a time lest we miss an artifact packed in red ochre that was difficult to see in a screen full of red rubble. The majority of the material was sorted through 1/4 inch mesh screens. Initially we ran several samples through very small mesh water screens. Water screening proved to be unproductive in that sometimes we would recover but a single small, shatter flake in three full screens of matrix. Partially because it produced so little, and partially due to the fact that the talus was a mixed deposit, we soon abandoned the water screens (Figure 3-7.).

What we immediately began to recover from screening was absolutely amazing, or a better term might be mind-boggling. By the end of the seven weeks that we worked in 2014, we had screened only 1/3 of the talus. During that time we recovered over 750 artifacts not counting common flakes. Artifacts included Sunrise, Clovis-era, Goshen, Midland, Folsom, Plainview, Agate Basin, Hell Gap, Cody and Frontier projectile points, along with numerous point preforms, gravers, spoke shaves and numerous scrapers and flake tools. We recovered additional artifacts that same fall when a crew of artifact collectors that are sponsoring research at Powars II, along with other volunteers, screened for a few days. More importantly, the collection from the talus revealed a great deal about the site without disturbing the deposits that remain intact on the hillside above. True, artifacts in the talus were recovered from a mixed context, but there is no doubt that they originated from the deposits in the site above, and were not mixed with materials from other sites, or that they represented activities unrelated to the ochre mine. As indicated by the large number of artifacts present in the talus, it seemed certain that the remaining intact deposit would also contain numerous artifacts along with other data. We could not have wished for a more productive field season. It went without saying that the remainder of the talus should be screened. Thus, we did so in the 2015 through 2019 field seasons with similar results each year. We eventually laboriously screened over 200 metric tons of talus and scree.

Based on results of the test excavation in 1986, the amount of red ochre present in that test, in the talus, and scattered all around the area, we were and still are confident that the site is a Paleoindian red ochre mine. There are numerous historic accounts of the area being littered with Indian artifacts, suggesting extensive use of the iron ore prehistorically for "paint" (Chance 1901, Snow 1895, Ball 1906, Bass 1962, *The Iron Gazette* 1899). However, from the variety of artifact types that we recovered from the talus, numerous questions arose with regard to what kinds of activities had taken place at the site. From investigations at other prehistoric mines (Frison 1991, Davis 1982, Reher 1991 and others) we expected the talus to contain mostly hammerstones and digging tools,

Figure 3-7. Some of the Expanding Environments crew at work screening the talus.

yet we found none in spite of careful examination of every rock and bone fragment. Moreover, we had recovered an inordinate number of projectile points along with very fine, delicate utility types of stone tools such as gravers and spokeshaves. What activities that assortment of artifact types represented in an ochre mine was perplexing. The collection from the talus yielded no clues as to how the mine functioned, or how the ochre had been extracted from the ore. Also, many of the projectile points and other tools are broken pieces, but we found no refits in this unusually large collection. It seemed probable, that the artifacts were used and broken elsewhere, and then deposited in the mine (Frison et al. 2018).

The projectile point types indicated that the mine functioned from Early Paleoindian through Late Paleoindian times, and became abandoned immediately after the Late Paleoindian era. That too was puzzling, since it would have been possible to obtain almost unlimited quantities of red pigment far beyond the indicated period of occupation, even up to recent times (Chance 1901). If later groups did utilize the mine, they did not leave culturally diagnostic artifacts behind as had the Paleoindian groups. Another problem was, the majority of the lithic artifacts were made from a material with which we were not familiar. That question was eventually resolved when we later discovered the source of that particular toolstone material near the ochre mine, a fact that turned out to be another intriguing and significant aspect of our investigations at Powars II.

One possible answer as to why Powars II was abandoned after the Paleoindian period, is that there are several accounts of an interesting feature that was discovered across the hills to the east, during development of what came to be called the CF&I Chicago Iron Mine. That feature was described as a large cave or crevice, dug into a deposit of soft or "paint" ore. Prehistoric use of the feature was evident by numerous artifacts that littered the floor of the feature. It was estimated that 250 tons of soft ore had been removed from the deposit by prehistoric workers (Weed 1927, Snow 1895, Rupp 1939, Chance 1901, Knight 1901, Gilder 1908). It seems likely that later cultural groups had simply located a more abundant and less labor intensive source of red ochre. It is also possible that the

Powars II source had been exhausted to the point that, after 3,000 years of use, mining became more difficult and less fruitful at that location. At any rate, we will never know for sure why Powars II was not used by later groups, or what transpired at the Chicago Mine location on the other side of the hill, because that site was totally destroyed by early historic mining activities.

One question that arose during screening the talus was one of projectile point typology. We almost immediately began to recover a large number of projectile points that we initially identified as Clovis, and they were also considerably more numerous than other point types. But the more of them that turned up (a total of 53), we began to realize that overall they varied from so called 'typical' Clovis points, of which there are many known from this general geographical area. As a group, they were short and basally thinned though not truly fluted, a type sometimes referred to as "Stubby Clovis" (Wormington 1965). When Vance Haynes visited the site in 2015 and viewed the points, he cautioned us that we should not call them Clovis, based mainly on the fact that they were not fluted (Haynes personal communication). For that reason, along with other point attributes, and following the suggestion of Bruce Bradley, we eventually began to refer to that point type as Sunrise points. In addition to the Sunrise points, a few Goshen, Folsom, Midland, Agate Basin, Hell Gap and Plainview points began to turn up. The Goshen points also differed from the 'typical' Goshen point type. The long unanswered question as to the true relationship between those Early Paleoindian cultural complexes, chronologically or otherwise, came to the forefront. It was obvious that we would not be able to address those kinds of research questions from the mixed deposits in the talus. A thorough analysis of the projectile points is presented later (Chapters 7. and 8. this volume).

We began relishing the thought that, perhaps stratified cultural deposits might be present in the intact deposits on the hillside above, or perhaps in a camping area located on the terrace on the valley floor below the ochre mine. Early European explorers reported approaching the area from the west up a rough trail through what became known as Eureka Canyon. At the upper end of the canyon they found a large, open, grassy valley that they named Sunrise Park (MacCannon 2003). Early historic photos depict small, historic mining camps located in the park on a large, gently sloping terrace in the valley below the ochre mine (Figure 3-8.). Several old photos also depict a deep channel located at the base of the hill directly below the site. Later photos show the CF&I mining town of Sunrise and industrial mining facilities covering the entire terrace; the channel and other low places were filled with mine spoils, and a rail line placed right up against the hill immediately below the ochre mine (Figure 3-9.). By the 1920s, the landscape on the floor of the valley was modified to a considerable degree by terracing, filling and other forms of land leveling to accommodate domestic and mine buildings, access roads and rail beds. A series of photos and maps indicate that over time, early buildings were replaced, and roads and other infrastructure relocated numerous times, apparently in response to evolving mining technology and domestic needs of the miners. All that activity completely obliterated the original topography of the valley.

Investigations in the Terrace

Nothing in the collection from screening the talus, or from a later test directly into the ochre mine, indicated that camping had taken place in or around the ochre mine, which is situated on a rather steep, rocky hillside. For example, evidence for domestic activity such as charred bones, burned and fractured rocks or charcoal, were not present in the talus or on the scree. As suggested earlier, because no refits for the large number of broken points and stone tools were found, it is doubtful that they were used and broken in or around the mine feature. Also, the paucity of shatter flakes suggested that very little tool manufacturing or maintenance activities were conducted in the mine. All evidence suggests that most, if not all of the lithic artifacts were manufactured and used elsewhere, possibly in a camp near the mine, and placed in the mine later.

The terrace in the valley below the mine looked to us like an ideal place for a prehistoric campsite. In spite of the extensive amount of historic disturbance, there was the possibility that intact, prehistoric deposits remained below the historic town and mine facilities. We decided to attempt some exploratory back hoe trenches in the terrace in

Figure 3-8. A broad, open terrace on the valley floor in 1902.

Figure 3-9. Rail line in front of and below site

search of prehistoric camps that might be associated with the ochre mine. We hoped to locate clearly stratified, dateable deposits indicative of the various periods of occupation represented in the talus. This in turn could address some of the questions that we had about the rather enigmatic artifact collection. During the 2014 field season we placed two short, back hoe test trenches immediately below the mine next to the hillside. We did not reach the bottom of historic fill in either test in that area even though our machine dug to a depth of 2.5m. We also tested several other places along the north edge of the valley with similar negative results.

In 2015 we tested two other areas on the terrace. One, the 2015 Terrace Test 1, was placed in the valley 50m to the south of the hillside directly across from the ochre mine. The upper 1.2m of that test is historic fill, consisting of heavily iron-stained mine spoils mixed with railroad and domestic trash. Directly below the fill we intercepted original, undisturbed terrace deposits. Using trowels and shovels, we excavated a 2m x 2m unit in the terrace deposits in arbitrary 15cm levels to a depth of 2.8m. The soil from the top to the bottom of the intact terrace deposits is a uniform, light gray or brown, sandy loam. A thin, dark level at 15cm below the historic fill contained a Late Prehistoric projectile point and several flakes. That level is lightly stained red from pigments that had leached down from the iron-laden fill just above. Occasional isolated flakes occurred between 15cm and 1m down from the top of the terrace. A rather obscure cultural level at a depth of 1m produced a Late Archaic point. At 1.15m, a level of a gravely, whitish chock rock contained a few flakes but no temporally diagnostic artifacts. A level heavily stained with red ochre at 1.6 m produced artifacts diagnostic of the Early Archaic period, including a side notched projectile point, a side notched knife, numerous flakes, a bone awl and several small bone fragments. Poorly defined deposits near the bottom of the unit yielded a projectile point of a type that we are not familiar, a small fragment of a biface midsection that exhibits evenly spaced collateral flakes along the edge, a few flakes and very fragmentary bone splinters, but no artifacts diagnostic of any particular cultural complex (Figure 3-10.). It became clear that undisturbed, intact deposits were present below the historic deposits.

Another test in 2015, Terrace Test 2, was placed on the valley floor 70m to the west and 60m southwest of the ochre mine. We soon discovered that we had located this test on the edge of an old, buried rail bed. Here, historic fill consisting largely of mine tailings and trash extends to a depth of 1.1m. Near the bottom of the fill we began to encounter numerous flakes and highly fractured nodules of chert and quartzite. We initially assumed that the breakage had occurred when the material was brought in from elsewhere for use as fill for the rail bed. But upon reaching the bottom of the historic fill, it became obvious that the fill actually rested on, or near the top of what appeared to be a natural outcrop of chert and quartzite. We then considered that the flakes were produced as the result of the top of the outcrop being crushed, bladed and pushed around by heavy construction equipment during leveling the grade for the rail bed. Upon closer examination, we realized that many of the nodules and flakes were multifaceted, and that many flakes exhibited striking platforms and bulbs of percussion. As we cleaned off the top of the outcrop, numerous cores and cortex flakes appeared along with several digging tools. The digging tools consist of chert flakes and core fragments that originated in the deposit (Figure 3-11.). Curiously, no hammerstones were represented. It soon became obvious that the outcrop was a toolstone quarry that was encroached upon, and damaged extensively by road construction, and that it consisted of a unique material that we had never recognized before. A highly interesting observation was that many of the stone artifacts recovered from the talus were made from this type of chert which is a very exquisite, high quality toolstone. We named the outcrop the John Voight (JV) toolstone quarry after the landowner, and assigned it site number 48PL2366 (Figure 3-12.). Later, we decided to use that site number to include the entire terrace on the valley floor in the town site.

The range of variation of the chert in the JV quarry is phenomenal. Colors vary from shades of red, brown, yellow, white, blue, green and nearly clear and translucent. Contemporary flint knapper D.C. Waldorf experimented with some of the material and came up with a variety of names for the variations in color and texture such as: White Chert, Red Cloud Agate, Tortoise Shell Agate and

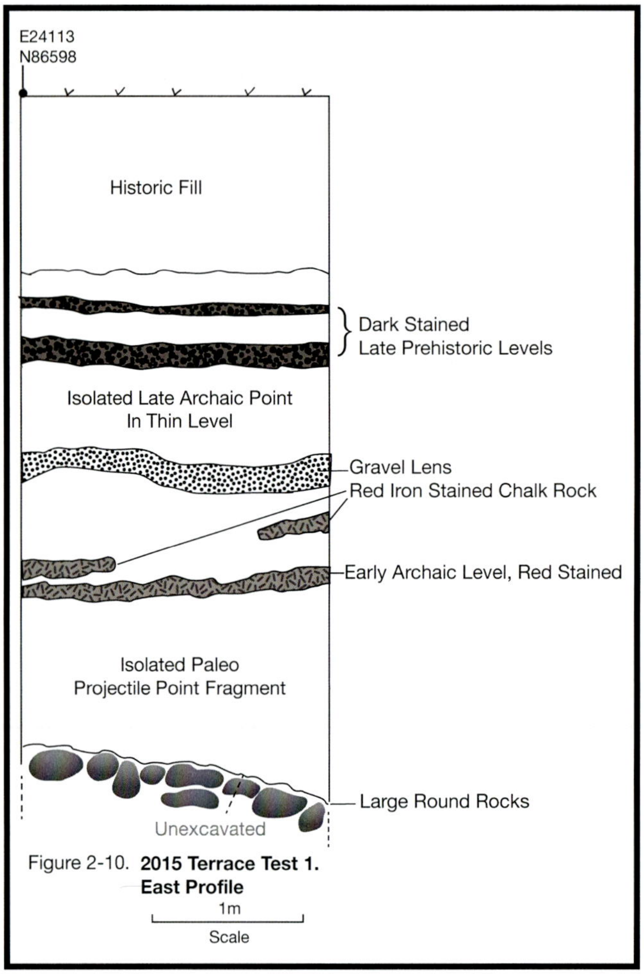

Figure 3-10. Profile of deep deposits located by a test excavation in the terrace in 2015.

Figure 3-11. Examples of digging tools from the toolstone quarry below the ochre mine.

Brown Chert and/or Chalcedony, Dark Reddish-Brown Jasper and Yellow Jasper (Waldorf 2016). He had not been provided with the full range of variation of the material to work with, or he could have expanded his description of the material to include more colors. Experienced rock collectors consider some of the chert, especially the chalcedony, and what is known as cream or plume agate, to be gem quality (Adams 2018). Colors and texture may vary considerably within an individual nodule. Some specimens contain dendrites similar to those found in classic Hartville Uplift material. Good quality chalcedony is also present. At least visually, some of the material is indistinguishable from Knife River Flint from North Dakota, and some is indistinguishable from Alibates found in Texas. Some of the chert fluoresces in ultraviolet light. Specific gravity is greater in this chert, and it is much heavier than classic Hartville Uplift or other material due to red streaks or veinlets of iron that run through some of the samples. Pockets or streaks of iron are also present in much of the quartzite in the deposit. The exterior of many of the nodules is heavily coated with kaolinite completely concealing the interior of the specimen. In order to learn what is inside the specimen, a knapper must chip away the koalinite cortex. The texture of the chert also varies considerably, ranging from a rather soft jasper, to chalcedony, to a somewhat grainy, agate-like material with occasional vugs of various sizes. The quartzite ranges from dark gray to a reddish brown, and the size of granules ranges from very fine to very coarse. However, quartzite is not nearly as abundant in the deposit as the chert. Some of the quartzite could, at least visually, easily be mistaken for Morrison Formation material from the Big Horn Basin. Both good quality chert and quartzite are sometimes present in the same nodule. The same range of variation in color and texture, is represented in the material of which the artifacts from the ochre mine are made. There is no doubt that the deposit buried on the valley floor is a Paleoindian toolstone quarry. It consists of a type of material not previously identified elsewhere, and it is directly related to the ochre mine on the hillside.

Figure 3-12. The top of the toolstone material immediately below historic mine deposits.

We pondered the genesis and geological context of this unusual type of toolstone. The rock structure at the south end of the Hartville Uplift is a complicated, convoluted series of faults and folds, of which the geological sequences and processes are not fully understood (Sims et al. 1996,1997, Chance 1901, also see Sutherland Chapter 6 this volume). Since we uncovered just a small portion of the top of the outcrop, and at that time had not probed into or around the deposit, we knew little about it. The materials in the deposit appears to be of Precambrian age, but it may be that the chert and quartzite deposit resulted from later geological events (Emmons and Eckel 1906). The toolstone is contained in a loose, grainy, yellowish or rusty colored matrix apparently formed in a sulfuric atmosphere. In addition to the chert and quartzite, the matrix includes inclusions of limonite, copper, malachite, chrysocola, azurite, kaolinite and goethite (Koski 2012). Some of the limonite is in mammillary form, is soft and pasty enough to have readily been used for yellow ochre, although we have no direct evidence for its use by the prehistoric miners. Ball (1906, pp 197) mentioned, "...great rounded bodies of limonite associated with iron-stained flint in the Sunrise open pit.". When heated, limonite can produce red ochre. We uncovered no evidence of limonite being heated, although it is possible that heating occurred sometime during the formation of the deposit (Wilkins 1967).

There is general agreement that this assortment of minerals forms in a sulfuric milieu, as is also the case for the iron ore, and it becomes distributed by hydrothermal processes into existing voids such as veins, fractures and pockets created by orogenic processes. These processes occur as a result of a series of compressional events in Early Proterozoic time (Sims 1997, Ball 1906, Rulenbeek 2007). While in solution, movement of the material can be up, down or lateral by ascending or descending currents, and deposits can be displaced by subsequent orogenic events that can amount to several hundred feet (Emmons 1901, Wray 1948, Wilkins 1967).

One hydrothermal process that generates all the various minerals in the JV quarry matrix, and that is also known to produce chert and quartzite, is a gossan. Gossans have been identified in several nearby locations in the Hartville Uplift area (Ball 1906, Snyder 1980, Hausel 1989). The JV deposit may not be a gossan, but it was definitely formed by a hydrothermal process, perhaps a gossaniferous process, and may be the remnant of a gossan that became dispersed by later orogenic events.

The question arose as to whether the exceptionally wide range of variation in color and texture in the JV chert could occur in one source, one possibility being that some of the material may have been brought in prehistorically from various quarry sources in other areas (Vance Haynes, personal communication 2015). Given the fact that the range of variation that most of the artifacts in the ochre mine are made of is the same as that in the JV deposit, then large amounts of raw material from other sources, would have to have been brought into the JV source by the prehistoric miners. Some of the large nodules of chert weigh as much as 150kg, and it is highly improbable that they were transported by prehistoric occupants (Figure 3-13.). And, a wide range of variation in color and texture is present within some of the huge, individual nodules. If the material was moved by historic mining, the loose, crumbly matrix described above surely would have fallen away during transport. None of the nodules or artifacts exhibit rounded edges produced from alluvial wear. An old photo of the mine depicts a rock outcrop in the area of the quarry, and a later photo depicts leveling of the outcrop by heavy machinery to prepare the area for a rail bed. The best evidence at the initial time of our investigation indicated that, the material in what we named the Voight quarry, rests in or very near its primary geological context, and that the chert and quartzite are associated with the iron ore and are Precambrian in age. However, a later test down into the deposit determined that some of the toolstone was displaced, but was not moved very far. It is likely that it was shoved aside to level the area. Therefore, in the future we will need to explore further beneath the rail bed to address the question of the primary source of the JV material. Still, the presence of the artifacts in the deposit leaves no doubt that quarry activities are represented, even though much of the breakage is definitely machinery related.

Further evidence that the toolstone and iron ore are related, is a large iron ore deposit exposed on a hillside in the old Chicago mine, located on the other side of the hill to the east of the Sunrise mine. There are also several small pockets of iron ore exposed in the road cut between the mines. That large exposure, and some of the small ore pockets, are

Figure 3-13. Geri Zeimens next to a large chert nodule in the John Voight toolstone source

encased in a yellowish, rusty matrix similar to the matrix that the JV toolstone rests in, and they also contain various amounts of the same chert and quartzite as the JV quarry. There is another abandoned iron mine, the Good Fortune mine, located over the hill less than 1km southeast of Sunrise. That mine was almost totally covered over by mine reclamation, but scattered around the area in proximity to the mine, are small nodules and flakes of chert similar to the toolstone at Sunrise. Paint quality ore or "Earthy Hematite" was also present at the Good Fortune mine, but in what context is not known (Sims et al. 1997). All the evidence indicates that the toolstone is associated with, and is the same age as the iron ore.

In an effort to better understand the genesis of both the toolstone and the red ochre, Frison asked geologist Wayne Sutherland to examine the deposits. Wayne prepared a detailed analysis of the geology of the area and of the site deposits (see Chapter 6 this volume). However, further research will be needed to determine the primary context of the toolstone. Petrographic, chemical and mineralogical assessments to obtain a signature for the JV chert, and how it differs from Hartville Uplift and other regional sources are needed. As discussed earlier, such studies will necessarily involve a large scale and long-term effort given the fact that, there are literally thousands of quarries located in the uplift in numerous formations of various ages.

Another test in the terrace in 2016, 2016 Terrace Test 1, produced some interesting results with regard to processing of the ore to obtain red ochre. This was a 2m x 2m test unit placed 5m west of the JV quarry, and also adjacent to the old, buried railroad bed. Here, the historic fill is 1m deep, and the top of the terrace immediately below the fill is stained a bright red. The red stain continues down to 20cm below the fill to a discrete cultural level with a hard-packed floor. The level contained several small flakes and one biface fragment along with several small bone fragments, but no temporally diagnostic artifacts. The level contained a shallow, circular pit that measured 50cm in diameter and 10cm in depth. This feature was completely filled with small, broken pieces of paint iron. The entire feature and the contents were heavily stained with red ochre. There was no evidence for burning in or around the pit, and after water screening, nothing other than small fragments of pulverized paint iron was found in the pit contents. The pit appears to have been filled but not reopened. There is no doubt that it is somehow related to processing iron ore to produce red ochre, but precisely how it functioned remains problematical. Due to time constraints, no further investigations have been conducted in this area (Figures 3-14., 3-15.).

Historic photos show that a broad, flat terrace originally existed on the north side of the valley, in the area directly in front of the main historic copper and iron ore mines (Figure 3-16.). That area is within 100m of the ochre mine and next to what is now a huge, deep, mine pit known as the Glory Hole. During reclamation of the mine in the 1980s, several meters of mine spoils were mounded on top of that terrace. The main purpose for that huge mound was to prevent water from flowing over the edge of the mine pit, and flooding the town of Hartville downstream at the mouth of the canyon. Because that mound is so large, access to that now deeply buried terrace in order to examine the old deposits will be extremely difficult. That situation is frustrating since that area of the terrace, based on examination of old photographs, appears to be one of the most likely areas to contain evidence for a prehistoric camp.

Whereas our search in the terrace was to locate prehistoric deposits, we soon became aware that we could not ignore historic archeological remains that we encountered. An example of a significant historic feature that appeared near the surface of one of our tests consists of a large, buried historic dump. We excavated and recorded a sample of the feature, and much of it remains intact. It contains quantities of broken glass, coal, nails, rusted cans, complete bottles, parts of iron stoves, shoes, and other pieces of clothing in an ashy matrix. Another example of a historic feature is a retaining wall discovered buried beneath .5m of fill in the vicinity of the JV toolstone quarry. We mention these findings here because, there is a great deal of potential for historical archeological research to explore poorly, or undocumented portions of the chronological history of the mine, the mining camp, and the lives of the miners and families that lived and worked on the terrace (Figure 3-17.).

Because most of the original landscape in the valley had become so obscure, Vance Haynes told us that what was needed was to cut a long trench from one side of the valley to the other. By doing so, we might be able to learn something about the geomorphic history of the valley, and to better identify areas where camps might be located. In that regard, in 2017 we hired a large track hoe excavator to dig just such an exploratory trench. The machine had the capability to dig a test 1.75m wide to a depth of a little over 11m. We began digging the trench in the valley floor next to the hillside directly below the ochre mine. We were surprised when at the starting point, just below the ochre mine, we dug to a depth of the limits of the machine and had not reached the bottom of historic fill. The fill consisted entirely of mine tailings mixed with trash. We continued the trench digging at full depth, and did not reach the bottom of fill until we had extended the trench 60m to the south. At that point the trench intersected terrace deposits at the depth of 11m. We found that the terrace slopes upward to the south at a 45° angle until at 70m to the south, it levels off to a more gentle slope, and from that point to the end of the trench, the terrace deposits are buried beneath 2m to 1m of historic fill. We continued digging to full depth through the fill and the terrace. We did not reach the bottom of the terrace deposits at any point. We terminated the trench at 80m from the starting point, when we intersected a drainage ditch and power line that were in the path of the trench (Figure 3-18.).

The terrace soils extend to an unknown depth below the bottom of the track hoe trench. The terrace deposits consist entirely of a light gray, sandy loam. We did not identify any unconformities in the terrace soil profile. However, we were not able to closely examine the trench walls because the loosely consolidated, historic fill kept caving in the sides, and for anyone to descend into that deep trench would have been extremely dangerous.

One important detail that we did learn about the valley from the deep test trench is that, at the time of historic contact, a deep, natural cut or draw

Figure 3-14. The top of a pit in the ochre-stained level in the 2016 Terrace Test adjacent to the John Voight quarry.

Figure 3-15. The pit in the 2016 test after excavation.

existed along the north edge of the valley. The original depth a point directly below the edge of the ochre mine would have been more than 20m. Over centuries, a good deal of deposits from the ochre mine, along with cultural remains that likely rested at the edge of the draw, would have eroded and fallen far below the current level of the valley floor and undoubtedly would have washed downstream. Consequently, we will never know the original lateral extent of the deposits along the southern edge of the site.

In 2021 we began a 1m x 2m test in the terrace 30m south of the JV toolstone quarry. It is adjacent to the old railroad depot, and the upper .75m of the deposits consist of debris from construction of the depot mixed with numerous flakes, but no culturally diagnostic prehistoric artifacts. From that level to a depth of 2m, the deposits are a gray, sandy loam and are culturally sterile. At 2m from the surface to a depth of 2.9m, the deposits contain a series of cultural levels that contain numerous flakes, and what appeared to be small flecks of charcoal. A small, circular pit feature that contained flakes was also present (Figure 3-19.). It was late in the season so we had to terminate our work without a thorough examination of the context of the cultural materials. During the 2022 field season we reopened the unit and collected what appeared to be charcoal, and submitted several samples for dating. The radiocarbon lab informed us that the samples appeared to contain organic material, but not charcoal. We had the lab date the samples anyway, which in essence meant that the dates would be from sediments rather than cultural material. The results were: 12,060 ± 50 BP with a calibrated range of 14,051 - 13,804 BP (Beta 644711), and 11,820 ± 40 BP with a calibrated range of 13,783 - 13,593 BP (Beta 645063). We also submitted an optically-stimulated luminescence (OSL) date from the same sediments,

Figure 3-16. An 1887 photo shows the terrace in front of the Glory Hole that is now buried below several meters of tailings.

Figure 3-17. East profile of an historic mine dump pit in the terrace.

and it yielded a date of 14,170 ± 560 (University of Baylor Lab. No. 5449). The 2022 test unit was small, only 1m x 2m. In 2023 we expanded the unit another 1m x 2m. Again, we uncovered numerous artifacts, mainly flakes, one end scraper and one broken biface, but nothing culturally diagnostic. We obtained four radiocarbon dates from the test in 2023 and again, they are from sediments rather than bona fide cultural remains and are: 11,700 ± 50 BP calibrated to 13,615 - 13,455 BP (Beta 671430), 12,050 ± 40 BP calibrated to 14,046 -13,806 BP (Beta 672466), 12,550 ± 40 BP calibrated to 15,111 - 14,806 BP (Beta 672465), and 11,680 BP ± 40 BP calibrated to 13,602 - 12,460 BP (Beta 672464). We also obtained two OSL dates from the same levels as the radiocarbon dates that are: 16,130 ± 520 BP (Baylor Lab. No. 5649), and 11,855 ± 470 BP (Baylor Lab. No. 5648). These dates are extremely interesting and exciting since some are older than Clovis. The fact that they also appear to be nicely stratified is also encouraging. George Frison said early on during our investigations that, "If there ever was a site that has the potential to produce evidence for Pre-Clovis, Powars II is it.". So far our test in those deep deposits consists of a small, $4m^2$ test unit. We are certainly looking forward to expanding the excavation next year, as it is critical that we determine the age of the cultural remains with respect to the age of the sediments.

Our exploratory test excavations have verified our suspicions that intact, stratified cultural deposits exist below the historic fill and mine facilities that now cover the valley floor. The problem is, the floor of the valley includes a vast area, some one hundred or so hectares, where original landforms are no longer visible on the surface (Figure 3-20.). The prehistoric camps that we are searching for are "like looking for a needle in a haystack", and the long, deep track hoe trench did not help much. Another deep trench placed in another area might be beneficial. Our goal is to locate discrete, datable cultural areas of activity affiliated with the various Paleoindian cultural complexes, identified by projectile point types in the talus and ochre mine. Results from these initial tests confirmed that the potential to do so is indeed a reality.

Dating The Site

For the most part, the age of the site deposits has been derived from relative dates based on projectile points representative of Paleoindian cultural complexes. The exception is two radiocarbon dates obtained from bone rod fragments from the talus, that are of the same time period as the Hell Gap cultural complex (Figure 3-21.) (Frison et al. 2018). The numerous radiocarbon and OSL dates have served to determine that, the site deposits are indeed of Paleoindian age. Thirteen dates from the interior of the ochre mine feature are from mixed deposits, and are not directly associated with any of the various cultural complexes represented by the projectile points from the site (Pelton et al. 2022).

A cultural level at a depth of 1.6m in the 2015 Terrace Test produced a projectile point and other artifacts diagnostic of the Early Archaic period. Projectile points above that level are diagnostic of Middle Archaic and Late Prehistoric period deposits, but were not recovered in association with materials suitable for radiocarbon dating. There is no evidence that any of the Archaic and Late Prehistoric cultural remains in the terrace are associated with the ochre mine. We have not determined with certainty the age of the earliest occupation. The old dates from the deep test on the terrace next to the railroad depot are intriguing, some are older than Clovis, and we need to learn if the associated artifacts are the same age as those old deposits. Based on projectile point types, the latest occupation of the ochre mine is represented by the Frontier Complex.

Reliable dates from contexts in unquestionable association with cultural remains, will be critical to future interpretations of data from the various aspects of the site. As stated above, for that to be achieved, well-separated, discretely stratified deposits will be needed. The potential for those kinds of deposits to exit in the terrace is high. Several highly reliable dates will be necessary to learn the age of what we are calling the Sunrise projectile point type and how it relates to other early Paleoindian complexes. However, one problem that we may encounter, is that since the points and tools were placed up in the ochre mine, including broken pieces, we may find that no projectile points were left in the campsite below the mine.

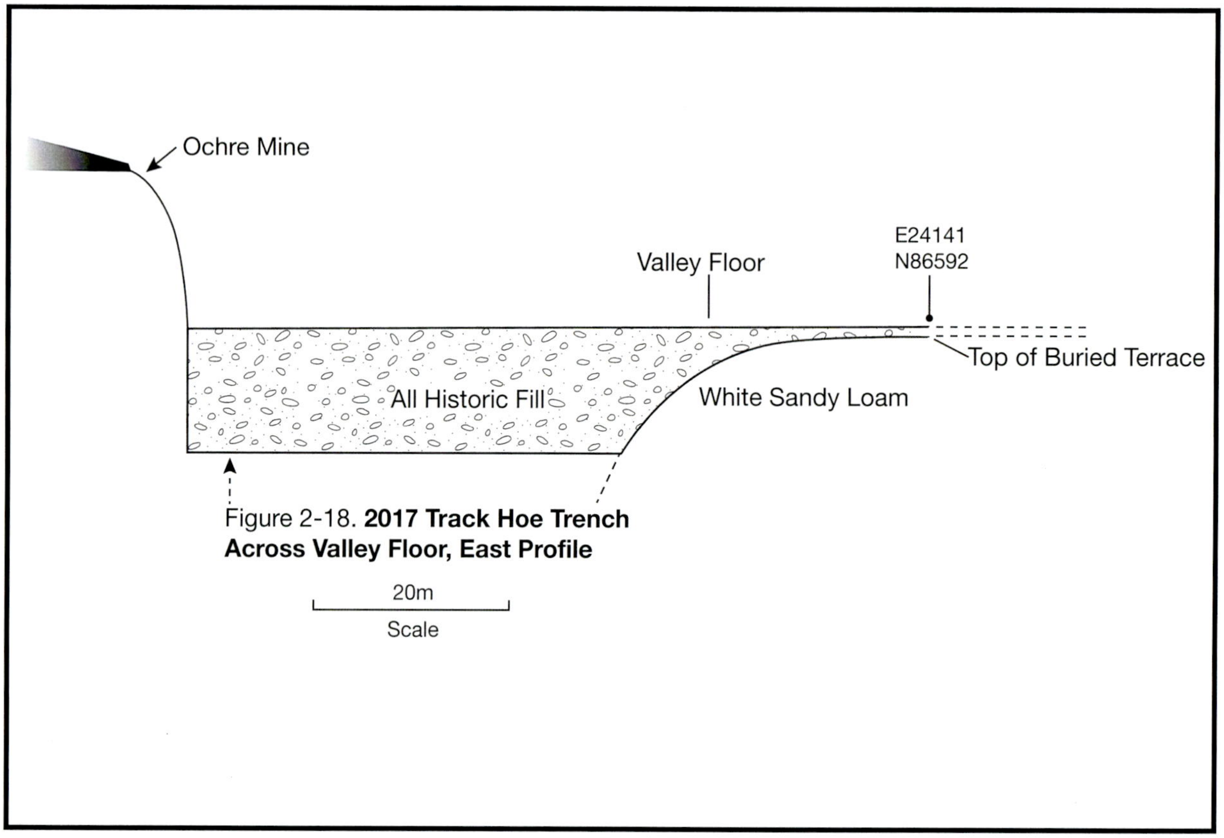

Figure 3-18. Profile of the long, deep track hoe trench.

Figure 3-19. Looking down into terrace test TT-22-1

Figure 3-20. Houses and mining facilities covering the entire terrace on the valley floor in 1920.

Figure 3-21. Bone rod fragments recovered from the talus.

Chapter 4

Journey of the Riser Cache and Renewed Interest in the Powars II Site

Steven L. Boles

This chapter is devoted to a group of 13 Paleoindian artifacts discovered over a hundred years ago by Dr. Frederick L. Riser, a mining camp doctor. (Figure 4-1.) The artifacts were initially thought to have been found near Rawlins, Wyoming, but we now believe it more probable that they came from the Powars II site. Briefly, the Powars II site represents a special use site where Paleoindians were mining red ochre. The late George Frison, one of the principal investigators at the Powars II site, felt that the extraction of red ochre and the discarding of weaponry at this locale were suggestive of ritual activity associated with hunting (Frison et al. 2008).

In 2018, Frison asked if I would write a chapter about the doctor's group of artifacts referred to as the Riser-Healy Cache in a 2008 co-authored Plains Conference paper (Boles et al. 2008). I accepted his offer and with what follows is part discovery story, part detective story, and part chance. Many distant and unrelated things came about that, in essence, according to Frison, rekindled his interest in the Powars II site and thus, along with other fortuitous factors in 2014, resulted in the ensuing work discussed in this volume.

The journey begins with how the cache came to light, a brief account on the discovery of the cache, research that provided details about the finder of the cache, and finally how the cache relates to Paleoindian material recovered at the Powars II site.

Dr. Riser's Paleoindian Collection

Howard Young, a retired barber and relic collector, brought the lithic material referred to as the Riser Cache to Southern Illinois in the summer of 2006. He was dispersing his artifact collection and had contacted a distant relative and fellow collector, Steve Healy of McLeansboro. Healy and I had recently become acquainted, and he contacted me and asked if I would accompany him to look over the Wyoming material. The cache included seven points, a point preform, two unifacial spokeshaves, and three large bifacial knives (Figure 4-2.). All the material was coated in red ochre, and

Figure 4-1. Dr. Riser (*The Rawlins Republican* Newspaper, 1902).

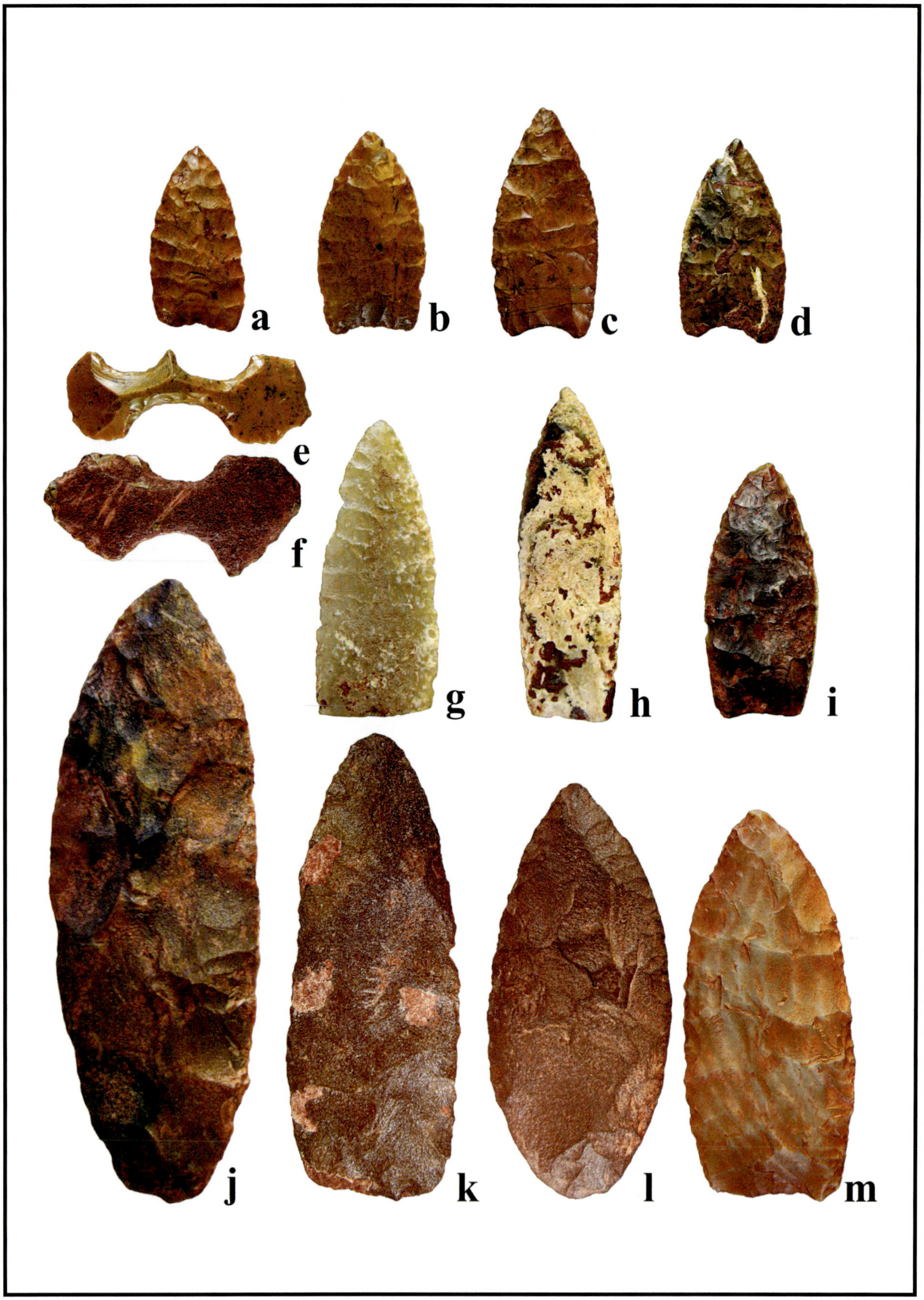

Figure 4-2. The Riser Cache.

the points were broadly similar to Paleoindian Clovis or Plainview types.

According to Howard, via Riser's family, the artifacts were reportedly found together by Dr. Frederick L. Riser sometime near the turn of the 20th century in Wyoming. We assumed the group represented a Paleoindian cache based in part on the family's account, but more importantly by the red ochre deposits on the artifacts reminiscent of the Anzick burial cache (Taylor 1969). Healy began researching Dr. Riser and gained additional information from Young that inferred that Riser may have found the cache when he lived in Rawlins, Wyoming. This locale for the group of material seemed quite plausible as the area contained large red ochre deposits as well as the discovery in the late 1800s of a butchered Mammoth by railway workers at site 48CR182 (Berner 2007).

Sometime prior to Dr. Riser's passing in 1938, the group of artifacts was passed down to his son, Frederick O. Riser. Fred O. passed away in August of 1987; the cache was then in possession of his widow. She later remarried and moved to Fort Collins, Colorado. Shortly thereafter, she began to settle her estate in the early 1990s. Her next door neighbor turned out to be George Frison's cousin, Robert Frison, who told her second husband to take the material to Young, with whom he had surface collected. The husband brought the material in a paper bag to Young's barbershop and informed him of the limited details concerning the discovery of the points and tools. Young kept the cache until 2006, when Healy acquired it.

Healy's interest in artifacts went beyond collecting, as he was also equally intrigued by archaeological study. Knowing of the cache's potential as archaeological importance, Healy offered the material to me for study. I was at the time in the process of obtaining my MA degree in Anthropology. I informed Healy that I would be glad to study it but really knew very little about Plains Paleoindians. The points in the Riser group were Clovis-like but not fully typical for the type (Frison et al. 2018). I also informed him that I had heard of Dr. George Frison and that many considered him the utmost authority on the subject, and further, that I would try contacting him.

Road Trip to Meet George Frison

After borrowing the Riser Cache, I found a number for George Frison and phoned him. George proved to be a pleasure to talk with, and he expressed an interest in seeing the artifacts. I informed him that I had a few days before my fall semester started and asked if I could bring them out for his inspection. Long conversation short, the next morning I was on the road driving to Laramie, Wyoming. The following morning I met George at his office in the anthropology building at the University of Wyoming. George inspected the Riser cache noting the similarities between Riser's group and material that had been collected at the Powars II site near Hartville, Wyoming. He retrieved the Powars II material and we compared it to the Riser Cache. Points from both were identical in point typology and the heavy coating of red ochre.

The Powars II site was located at the edge of a defunct century-old (1881-1980) iron mine known as the Sunrise Mine. However, given the reported possible find spot at Rawlins, and with a red ochre deposit in the same area, it was impossible to refrain from questioning the context of Dr. Riser's artifacts, and why they were so similar to the Powars II material. After a morning of discussing Paleoindian material, George offered to show me the Hell Gap site located a couple hours north of Laramie. Before getting to Hell Gap, George stopped in Hartville at a gated road. He stated that the Powars II site was just a mile up the road but this was as far as we could go as the landowner was not a fan of archeology. After a tour of the Hell Gap site, we did not get back to Laramie until late in the evening.

Before I departed the next day, George suggested that we do more research on the cache and also see whether we could source the ochre on the Riser group. Upon returning to Illinois, I contacted staff at the University of Missouri at Columbia in regards to sourcing the red ochre. They felt it was possible, but doing so required destructive analysis to the artifacts, something that Healy was not in favor of, thus, that line of study would not be pursued.

I also began searching for details about Dr. Riser hoping that I might come across an old newspaper account of his discovery of the cache. George Zeimens, life-long colleague of Frison and co-principal investigator at Powars II, later joined in this effort. Together we were able to piece together a fair background on Dr. Riser, who was born in Lansing, Iowa in 1858. Although we did not come across an account of his discovery of the Paleoindian cache, a short write-up in the semi-weekly *Republican* newspaper from October 1902 stated that near a creek south of Rawlins, Dr. Riser found an elk antler tool with a piece of iron attached to one tine for use in scraping hides. It seemed possible that his family had mistakenly assumed Riser's find along the creek to be the provenance of the cache.

We discovered that Dr. Riser was quite wanderlust. After graduating from the German-English College in Galena, Illinois, he taught school for four terms in northeastern Iowa. He then went to medical school in St. Louis, graduating in 1884. From there he practiced medicine in several western towns. He was also an avid bird-watcher and reported a Caspian Tern in the spring of 1893 near Lincoln, Nebraska. He moved to Douglas, Wyoming from Lincoln in 1897; from Douglas to Hartville in 1899, and according to the Iron Gazette, built two houses on his lot there. Not long after, he moved to Rawlins in 1902 and practiced in the Hoffman Building with partner M. L. Hooper. While in Rawlins, his abilities and popularity was rewarded by the board of commissioners with an appointment as county physician. His last known residence was in Irondale, Colorado, according to the 1920 U.S. census.

Use-Wear Study and Healy's Passing

Frison felt that a use-wear study would be of value and suggested I contact Marvin Kay at the University of Arkansas in Fayetteville. I met with Kay in February of 2008, and after examining the artifacts, he agreed to do a use-wear analysis on the artifacts in the cache. Collectively, we planned to co-author a presentation on our research of the Riser artifacts for the fall Plains Conference in Laramie, Wyoming. Healy was excited by the prospects of our research but unfortunately, did not get to see the results of our study as he passed away on March 19, 2008. Healy's family graciously allowed Kay to complete his study.

Kay's use-wear study is similar to ones he conducted on the Colby site artifacts and on Goshen and Folsom artifacts from various sites (Kay 2008, 1996 Root et al, 1999). Relative to the Riser-Healy collection, Kay determined that all of the items functioned as tools and that tool function largely followed tool form. All seven chipped stone points were consistent in being manufactured as projectiles, and then used secondarily, or in an ad hoc fashion, as knives. This record of use is identical to that of the four Colby points and more generally is the norm for Clovis. Of the four larger bifaces, two are ovoid and two are rectanguloid (Figure 4-2. j-m). The smallest rectanguloid specimen of the concentrically banded chert appeared to have been used first as a knife and then as a projectile point, which makes it unusual and without a formal analogue in early Paleoindian assemblages (Figure 4-2. m). This tool has diagonal percussion flakes, some of which could be characterized as Clovis-style overshot flakes along with pressure retouch. The other three all were likely hand-held butchering knives and fall formally into two groups. The first is that of likely Clovis-type knives; the two largest knives in the cache both exhibit Clovis technology with large percussion flakes and little pressure retouch (Figure 4-2. i-k). The second grouping is that of an ultra-thin normally associated with Folsom and is represented by the smallest of the three knives (Figure 4-2. i). That knife exhibits Folsom ultra-thin technology wherein the percussion flakes were executed in such a way, that each flake dove into the center, making the center of the knife thinner than the outer edge. In this instance, the specimen is thicker than Folsom ultra-thins, but shares technological attributes. Its pointed end is burin facetted, but this is likely to be a byproduct of striking a hard surface such as a bone rather than an intentional preparation, and is without burin use. It also exhibits diagonal percussion flake scars and pressure flaked edges due to re-sharpening.

The final two artifacts are rather unusual unifaces; both functioned as spokeshaves and as slotting burins (Figure 4-2. e,f). Both double crescent, unifacial spokeshaves are similar to a spokeshave found in Folsom contexts reported by Hofman and Graham (Hofman & Graham 1998:99-100) and the Lindenmeier site (Wilmsen and Roberts 1978). Though spoke shaves are occasionally found in Paleoindian tool kits (Frison & Bradley 1999:34), one of the spokeshaves from this cache is unique in design with a graver tip in the center of one of the crescent's depression (Figure 4-2. e). Either or both could have been used in the production of wooden shafts or bone rods.

Paleoindian Conference at University of Illinois-Urbana

In April of 2008 I attended a Paleoindian conference hosted by the Illinois Transportation Archeological Research Program (now known as the Illinois State Archeological Survey) at the University of Illinois in Champaign-Urbana. As mentioned earlier, the Riser points are loosely similar to Clovis and/or Plainview types, but are much stubbier than both point types, and flute-like flakes are very short and appear more as single basal thinning flakes than a true channel flakes. I had hoped to get additional feedback from participants on point type as well as raw material types.

Most everyone who examined the Riser material felt that two of the seven points were morphologically consistent with the Plainview type and both appeared to be made from heavily patinated Knife River Flint from North Dakota (Figure 4-2 h-i). The four remaining complete points were all thought to be made from chert that most felt was very similar to dendritic Hartville Uplift chert in Wyoming (Figure 4-2 a-d). However, no one could conclusively identify a specific point type for the stubby or squat "Clovis-like" points. Other point types offered for the stubby points included Colby and Jimmy Allen. The seventh point was a distal portion and made from a heavily patinated chalcedony, some suggested Flattop Butte chalcedony from the border area of Colorado, Wyoming, and Nebraska (Figure 4-2. g). The cache also includes a late stage point preform made from a heavily patinated blue and white banded semi-translucent chert that some thought might be banded Alibates from the Texas Panhandle region (Figure 4-2. m). Two ovoid knife blades are of a quartzite that some suggested resembled Spanish Diggings quartzite. The third was thought by some to possibly be Niobrara jasper. The remaining two artifacts in the group are both unifacial spokeshaves, one with a graver spur in the center of one notch, and both appear to be made from Hartville Uplift chert identical to that used for the four squat points.

Plains Conference Paper

I presented the culmination of our initial research at a symposium held in honor of George Frison at the 66th annual Plains Conference in Laramie, Wyoming in October 2008 (Boles 2008).

Table 1. Dr. Riser Artifact Group

artifact type	*material*	*Length cm*	*Width cm*	*weight g.*	*Figure 1*
projectile point	Hartville	4.6	2.5	6.7	a
projectile point	Hartville	5	2.7	10.7	b
projectile point	Hartville	5.7	2.5	9.4	c
projectile point	Hartville	5	2.6	12.2	d
spokeshave/graver	Hartville	6.2	2.2	6	e
spokeshave	Hartville	6.5	3	15.3	f
projectile point	chalcedony	7.2	2.9	15.2	g
projectile point	un-named	8.2	2.4	12.6	h
projectile point	un-named	6.3	2.8	12.8	i
ovoid knife	quartzite	15.6	5.4	117.4	j
knife blade	quartzite	11.7	4.4	62.2	k
ovoid knife	quartzite	10.2	4.4	43.2	l
preform	un-named	9.6	4.2	41.7	m

At that time we compared two of the points to the Goshen type and the stubby Clovis-like points to the Colby type. Frison later compared the latter to a Clovis-like variant of the Northern Plains (Frison et al. 2018). These points have subsequently been identified as a newly defined type referred to by Bruce Bradley in this volume as "Sunrise" points (see Chapters 7 and 8).

After the presentation, Frison introduced me to Dennis Stanford, Director of the Paleoindian/ Paleoecology Program at the National Museum of Natural History at the Smithsonian. The three of us discussed the close similarities of the Riser group with the material Wayne Powars had donated to the Smithsonian, as well as the material recovered by Frison's 1986 test trench at the Powars II site (Stafford et al. 2003). At that time, we had not discovered that Dr. Riser had practiced in Hartville, but we all agreed on the likelihood that Dr. Riser had found the material at the Sunrise site. Now that we know Dr. Riser did indeed live in Hartville, we feel there is little doubt that Dr. Riser found the 13 Paleoindian artifacts in the Powars II site area. Still open to further inquiry, is whether the 13 artifacts represent a cache, or merely a random collection of artifacts recovered from the vicinity of the site. One point of resolution that can now be offered is that, except for the specimens made of Knife River Flint, the various raw material types present in the Riser cache, including the point fragment made from chalcedony, can now be tied to a chert quarry located very close to the site (Chapter 3 this volume).

Acknowledgments – I will forever be grateful for the good friends I made while on the journey of researching the Dr. Riser cache. Some of these friends have passed on along the way; some of us are still hanging on. I'm indebted to the late Steve Healy, the late George Frison, and to George Zeimens. Without these fine gentlemen and scholars, the work presented here would not have occurred. I'm also grateful to Sherri, my wife of 40 years, who has been supportive and understanding of my varied archeological interest and related travels. I would also like to thank Patricia Wright, my wife Sherri, and George Zeimens for the comments and suggestions that improved this chapter. Any and all errors are mine.

Chapter 5

The Ochre Mine and Red Ochre

George M. Zeimens

The Ochre Mine Feature

The feature that delineates the precise arca where the prehistoric miners were digging, is located on a hillside on the north side of the valley directly across from the abandoned mining town of Sunrise. On the surface the feature consists of a shallow, linear, open trench or drift 10m in length, that extends into the hillside some 12m to 15m above the valley floor. The trench trends uphill on the detritus covered hillside in a northeast direction. It was excavated into an allogenic iron ore deposit situated on an approximately 20° slope of the hillside. To prepare a path for a railroad line, CF&I had cut deeply into the hillside directly below the site, which left the exposed site deposits perched on the edge of an almost vertical cliff. The south end of the trench is at the edge of the cliff, and it extends up the hillside from that point. Close examination of an early 1920s photograph of the mine, shows what appears to be the ochre mine trench on the hillside (Figure 5-1.).

Only 10m of the trench is visible on the surface, because the northeastern or upper end, remains covered with trash that slid downslope from the CF&I mine haul road that extended along the hillside above the site. We do not know just how far the trench extends uphill beneath the trash, but if it is more than about 4m, it was truncated in that area during preparation of the highwall at the western edge of the historic iron mine area called the Glory Hole. The southwest or lower end of the feature terminates on the edge of the cliff that falls almost straight down to the valley floor. We could not determine how much of that end of the feature was lost from erosion. We know that the southwest end at the edge of the cliff is where Wayne Powars dug for his collection, and we have no idea how extensive his digging was.

On the surface the trench is 2.5m wide. It is oriented at about a 30° angle to the east of the plane of the slope, so that the south edge of the trench is lower than the north edge. Both edges of the feature are lined by low, linear mounds of prehistoric mine tailings. The surface of the mound along the north edge is covered with what appears to be eolian dust partly derived from a loose, grainy material, apparently historic mine fines, that was pushed over the edge of the haul road above. A few rocks protrude from the top of that mound. The mound on the south or lower edge of the trench, is lined with a dense pile of nodules of blue iron ore the top of which is fully exposed on the surface. The nodules average 15-20cm in diameter. There can be no doubt that the linear pile of heavy, dense, blue iron nodules along the edge of the trench, is what prevented the mine feature from becoming lost entirely to erosion many years ago. Erosion extends right up to the southern and eastern edges of the site, and terminates right at the edge of the mound

Figure 5-1. The Ochre Mine Trench Feature is Visible in this old photograph

Figure 5-2. Blue iron nodules along the south edge of the trench feature

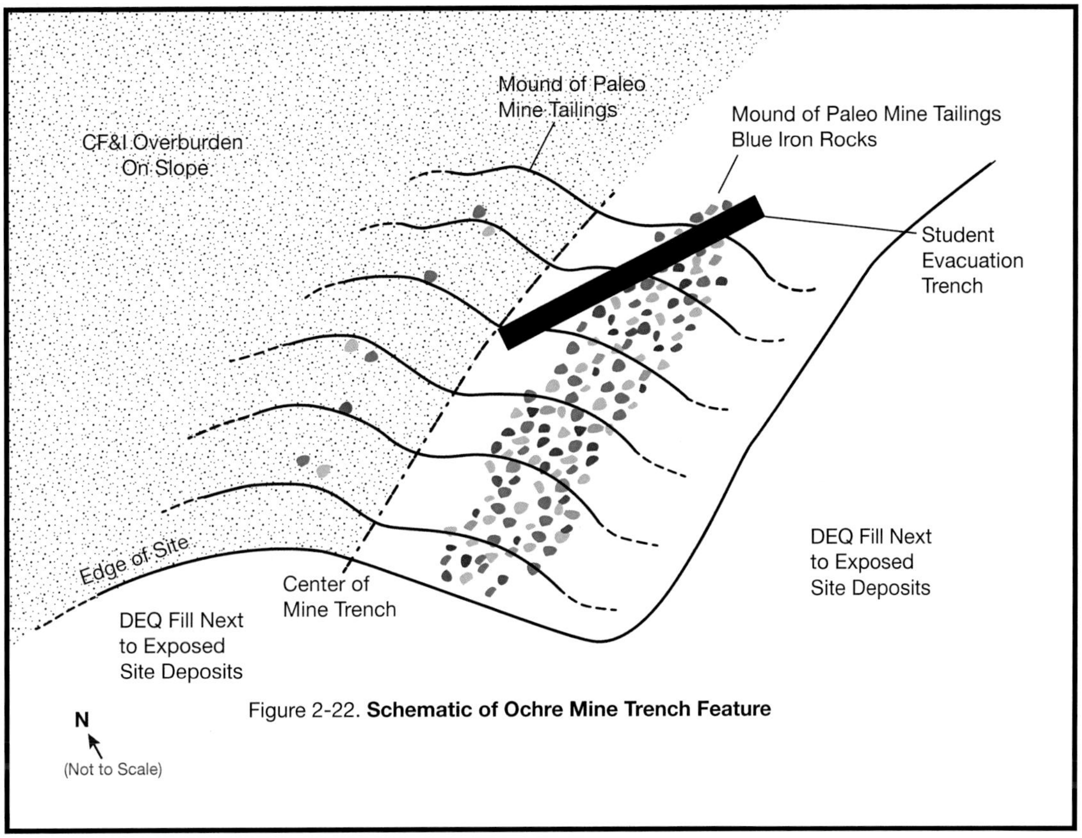

Figure 5-3. Schema of the ochre mine trench feature

of blue iron nodules. To the west of the trench, the site has eroded a short distance into the hillside (Figures 5-2.,5-3.).

The Figure 5-3. drawing of the trench included here does not serve as a viable record of the feature. A good profile needs to be produced by an excavation perpendicular to the long axis of the feature. A test excavation into the feature (discussed below) was oriented at an angle to the trench (Pelton et al. 2022). Unfortunately, the east end of that excavation removed the north end of the line of blue iron rocks. Also, during the test excavation, numerous nodules were removed from the top of the pile, and freshly excavated rocks were placed on top of the pile. These acts seriously damaged the integrity of this rare Paleoindian feature. The west portion of the excavation terminated at the center of the trench, so only a small portion center of the feature was included in the test. Therefore, a profile that delineates the width, depth and a cross section of the feature remains to be achieved, and would be an important aspect of the record of this rare Paleoindian feature.

The mineral deposit that was the focus for prehistoric mining activity contains three kinds of iron ore. The blue iron mentioned above is very dense, hard and heavy, and apparently was discarded when encountered by the miners. It is not easily broken and does not contain soft ore or ochre. The so-called "paint" ore is much lighter, softer, grainier and in many cases crumbly. It contains an abundance of soft, crumbly material called hematitic ochre or paint ore. Ochre is the result of decomposition of the ore which with Sunrise iron, may be partially due to bacterial activity (Tankersley et al. 1995). The bacteria colonies (prokaryotic bacteria) are ancient and are alive and active today. It is possible that the bacteria could be useful for tracing ochre from the Powars II source. A third type of ore in the deposit is an earthy material that appears to be a mixture of fine, powdery hematite bonded with illite clay, that has a consistency very similar to a crayon (see Sutherland Chapter 6 this volume). This admixture easily produces a bright red color when even lightly touched, and was found to be present in the form of a thin lens at the bottom of the deposits in the ochre

Figure 5-4. Two soft paint ore nodules can be knocked together to release ochre

mine feature. It is possible that this material may have been the most sought after type of ochre by the prehistoric miners.

We found no direct evidence for how the ore was processed to obtain ochre, how it was collected, or for how the mine functioned. However, while working through the matrix in the mine deposit during a test excavation (discussed below), we made observations that may offer some clues. When the Paleoindian miners encountered nodules of worthless heavy blue iron, they apparently were discarded by placing them along the southern edge of the trench. The north edge is higher than the south edge, and slopes toward the center of the trench. During mining, if the nodules were placed on the north edge, they might have fallen back into the trench and become a nuisance, or perhaps even a danger to the miners. That would explain the pile of blue nodules along only the lower edge of the trench.

The softer paint ore is instead porous, and we inadvertently discovered that, if two nodules are knocked together they crumble, and hematite dust or ochre literally pours out of the cracks (Figures 5-4., 5-5.). However, that method of extracting the ochre works only if the paint ore nodules are freshly extracted from the deposit. Because the paint ore is so porous, if a nodule is left in the open, rain water percolates through the cracks and washes the powdery ochre out and onto the ground. Thus, the need to dig, or to quarry into the iron deposit to obtain fresh nodules becomes obvious. The idea that knocking nodules together might have been used to extract ochre from them is bolstered by the fact that, the matrix inside of the mine feature consists mainly of small, pulverized chunks of the soft ore nodules. Based on tests in the area adjacent to, and outside of the mine feature, crumbled, pulverized pieces of the soft ore do not occur naturally in the deposit on the hillside. During screening of the matrix, the material that remained in the screen consisted almost entirely of small, broken pieces of paint ore (Figure 5-6.). Because the paint ore nodules are so crumbly, hammerstones would not have been needed to break them. Only two hammerstones were found in a test excavation in the trench and

Figure 5-5. Ochre produced after nodules knocked together

one in the talus, and they exhibit severe use in the form of large, wide flake scares on the working edges. Because paint ore is so soft, the amount of heavy use exhibited on the hammerstones could not have been produced by hitting them against the soft paint ore nodules. In that regard, it is possible that the hammerstones were used against a hard surface of some kind elsewhere, possibly toolstone, and for some reason brought up to the ochre mine.

In 2016 we were getting close to finishing screening the talus and we wanted to learn more about the nature of the intact deposits in the ochre mine. At age ninety-three Frison's legs were a little wobbly, and it would be too dangerous for him to work on the steep, rocky hillside. In an effort to involve students in our Powars II research project, he decided to offer a University of Wyoming graduate student the honor of working on this amazing site. Graduate student Spencer Pelton was chosen to place a test excavation directly into the ochre mine feature. He put together a crew of students and other volunteers, and marked out six, 1m x 1m units at an oblique angle to the linear

Figure 5-6. Residue in the screen from the matrix in the trench feature

orientation of the mine trench (Pelton et al. 2018). Pelton was asked to contribute to this volume but declined the invitation.

Because of the extensive amount of historic mining in the area, we initially thought that the trench was a historic prospecting pit. That turned out not to be the case, as the feature is entirely prehistoric. As we had expected, the mine deposits proved to be as rich as the talus and scree deposits. We also considered that intact deposits in the site might be mixed due to a sequence of later cultural groups digging through deposits that had been disturbed by earlier groups. That did turn out to be the case. However, Pelton does think that we were able to identify discrete, culturally diagnostic pockets of activity in the mine feature (Pelton et al. 2022). A small, prehistoric pit dug down through earlier deposits in the trench contained three Hell Gap projectile points. It is not known whether the pit is of Hell Gap age, or if Hell Gap miners tossed the points into an existing pit, or whether the points eroded into the pit from more recent upper deposits. The student test resulted in the recovery of 1,215 artifacts.

A majority of the test deposits were water screened. Water screening of numerous 12 liter (5 gallon) buckets of matrix, produced only a few shatter flakes. Many more shatter flakes would be expected if flint knapping was a major activity in the mine feature. The types of artifacts found in the test in the mine feature, did not vary significantly from those from the talus with the exception that, Hell Gap projectile points outnumbered other point types in the test, whereas Sunrise points greatly outnumbered all other point types from the talus. The difference in percentage of points types from the two deposits, can be attributed to the fact that the test included but a relatively small portion of the mine feature.

We succeeded in learning several things from the test excavation. First and possibly most importantly, we were able to verify that the site is entirely of Paleoindian age, and that the deposits in the mine feature are not contaminated with historic materials. Secondly, the site does not contain evidence for camping or other domestic activities such as charcoal, butchered bone or heated rocks. The main activity represented is mining for red ochre. It is possible that some flint knapping took place, but apparently not to any great extent. From bone and antler tools, we learned a little about how the ore may have been loosened from the matrix, and by experimentation, we learned how ochre could be extracted from the ore by hitting paint ore stones together. The precise length and width of the trench and other aspects of the feature remain to be determined.

We considered the possibility that stratified deposits might be present adjacent to the mine feature. Less than 1m of deposits remain at the outer edge of the mound of blue ore, and absolutely no deposits remain on the southwest end of the linear trench, which is at the edge of the cliff. However, we estimate that there are roughly 250m^2 of intact deposits on the hillside on the northwest side of the trench feature.

To determine the nature of the deposits adjacent to the feature, we placed two small test units, one west of the feature, and one above the feature. Both of them were found to contain culturally mixed deposits, and matrix in those tests did not contain pulverized paint ore like that present in the trench feature. However, a tightly packed pile of flakes at the bottom of the west test unit appears to represent a small, discrete, intact remnant of a cultural level. A small, very delicate graver was contained in the pile, and an overshot flake was located at the edge of the pile. Also, the base of a Sunrise projectile point rested next to the pile. All of the flakes, including the overshot flake, are of toolstone material that obviously originated in the nearby John Voight quarry. The pile appeared to rest in primary context as though it was intentionally placed or dumped at that location. If the pile was produced by knapping at that spot, the flakes most likely would be scattered over a wider area. Also, when the pile was water screened through a very fine screen, only 15 small, shatter flakes were found. It appeared that the knapping had taken place elsewhere, and the debitage intentionally transported to, and left beside the ochre mine (Figures 5-7., 5-8.). An OSL date obtained from the bottom of that test is: 11,490 ± 410 BP (Baylor Lab. No. 5448). Two bone fragments were radiocarbon dated and are: 9,860 ± 30 BP with a calibrated range of 11,322 - 11,205 BP. (Beta 873463), and 9,880 ± 30 BP with a calibrated range of 11,345 - 11,217 BP (Beta 672462). Those dates are not in direct association with diagnostic

Figure 5-7. Profile 1 from the OCM-21-1 test west of the trench feature

artifacts. Results of the tests determined that the deposits adjacent to the mine feature are also mixed, and that we do not yet know the boundaries of the site.

Red Ochre

Red ochre is an earthy red pigment. There are several materials that produce red pigment. The material at the Powars II site is known as earthy hematite, Fe_2O_3, anhydrous iron oxide, ferric oxide, red earth or paint ore. The prehistoric use of red ochre is a world-wide phenomenon that extends as far back in time as the beginnings of humanity. Red pigment has been a component of the human story for thousands of years, and may have actually been a factor in the social and cultural development of mankind (Wreschner 1980). Red ochre has been found in several sites associated with Neanderthal remains in Europe at 200-250ka (thousand) years ago. It has also been found at even older sites in Europe, Africa and Asia, including sites with evidence for the earliest Homo sapiens (Roebroeks et al. 1012). Red ochre was used extensively for one of the colors in the elaborate, ancient paintings in many of the caves in Europe (Rigaud 1988, Gourhan 1968).

Today, red ochre is used mainly as a rouge and for other cosmetic purposes, for artist's paint, and commercial paint for buildings. It can be purchased today at numerous commercial outlets including the internet. It is in use today for a variety of purposes by native people in the Americas, as well as natives in many parts of the world. Prehistorically, it is known to have been used for ornamentation of skin and other objects, in artwork, as a preservative and in mortuary situations (Roper 1991, Morrow 2016, Hodskiss 2013, Hovers et al. 2003). It has also been used as an adhesive, an abrasive, as a sunscreen and insect repellant, and even for medicinal purposes (Tarlach 2018, Velo 1984). There is a lady that lives in this area who uses ochre from the Powars II mine on her feet. I have not questioned her enough to find out whether she benefits physically or spiritually from the application.

In addition to functioning for many secular uses, there is strong evidence for spiritual and/or ritual use. The fact that red ochre is present in many instances in burial situations, is considered to be strong indication for spiritual involvement with the pigment (Morrow 2016 and others). We do not have much direct evidence for how Paleoindians used red ochre. It was present in the Paleoindian graves in the Anzick site burials in southern Montana (Lahren and Bonnichsen 1974), and in the late Paleoindian period Gordon Creek burial in northern Colorado (Breternitz et al. 1971). Red ochre is often found in association with caches of Paleoindian projectile points and other tools (Frison and Bradley 1999).

As discussed earlier, we do not understand why so many artifacts other than quarrying tools are present in the Powars II ochre mine. The only tools commonly found at prehistoric quarries or mines tend to be hammerstones and digging implements. We have recovered over 150 projectile points and point fragments, some of which were heavily plastered with red ochre, from Powars II. None of them show use for digging or scraping, and some of them exhibit the kinds of brakeage and wear from use in hunting (Frison et al. 2018). Of the fragments, there are no refits. There is also an equal number of very fine, delicate utility tools such as gravers and spokeshaves in the collection, and many of them are fragmentary. Refits were found for only three of the tools, and it appears that they were fractured after deposition. It is logical to assume that, if the artifacts had been broken during use in the mine, refits would be found. Also, the pile of flakes described above in a test unit near the mine feature, appeared to have been produced elsewhere and dumped in the site. If produced by a flint knapper at that location, the flakes would have been scattered over a wide area. That, along with fact that very few shatter flakes were present in the pile or in the mine trench feature, is an indication that knapping, manufacturing or resharpening of lithic implements were not major activities in the site. Numerous common flakes were present, but again, if they were produced in the site, more shatter flakes should have been recovered. A few core fragments were recovered, but they were small nodules and lacked cortex.

In view of the above, we have concluded that most of the lithic artifacts were manufactured and used elsewhere, and carried up the hill to the ochre mine and left as offerings. The purpose for the offerings is conjectural. George Frison suggested that these kinds of offerings, especially the large number of projectile points that appear to have been used, may be an indication for hunting ritual (Frison et al. 2018).

In the next chapter, Wayne Sutherland provides an excellent overview of the geology of the rocky, hilly area that encompasses the ochre mine. The genesis of the iron ore and ochre deposit is poorly understood at this point in time. Sutherland breaks down the mineral and chemical properties of the Powars II red ochre source, and compares it to a red ochre source about 80km to the west near Rawlins, Wyoming. The Rawlins source was formed by geological processes much different than that of the Powars source. Sutherland's description and mineral and chemical analyses of the red ochre will be useful for eventually determining a signature for the Powars II source, and how it differs from other sources. A recent study suggested that, based on properties present in the red ochre found at the Fetterman (La Prele) mammoth site, it is traceable to Powars II (Zarzycka et al. 2019). However, to reliably trace ochre to its source will not be possible, until numerous samples are analyzed from each source to determine if variation exits within each source, something that remains to be done (Dayet 2023, Tankersley et al. 1995). Having the ability to trace red ochre found at archeological sites would be a major contribution to studies of territorial dynamics and other areas of Paleoindian research (Brooks et al. 2018).

The red ochre at Powars II is of a very high quality. The color is of a bright, rather illuminant red. It is slightly oily, and not crumbly like some of the ochre we have seen from other sources. When only slightly dampened it becomes pasty. It spreads easily, penetrates deeply and weathers well. In fact, once applied it is hard to remove. For example, when CF&I sold their vehicles after the mine closed in 1980, many of them had red iron-stained tires. Some of the vehicles are still in use today in other areas, and the tires are still red. It is possible that prehistorically, the Powars II red ochre served as a trade item. If it did, it likely had a high value due to its high quality.

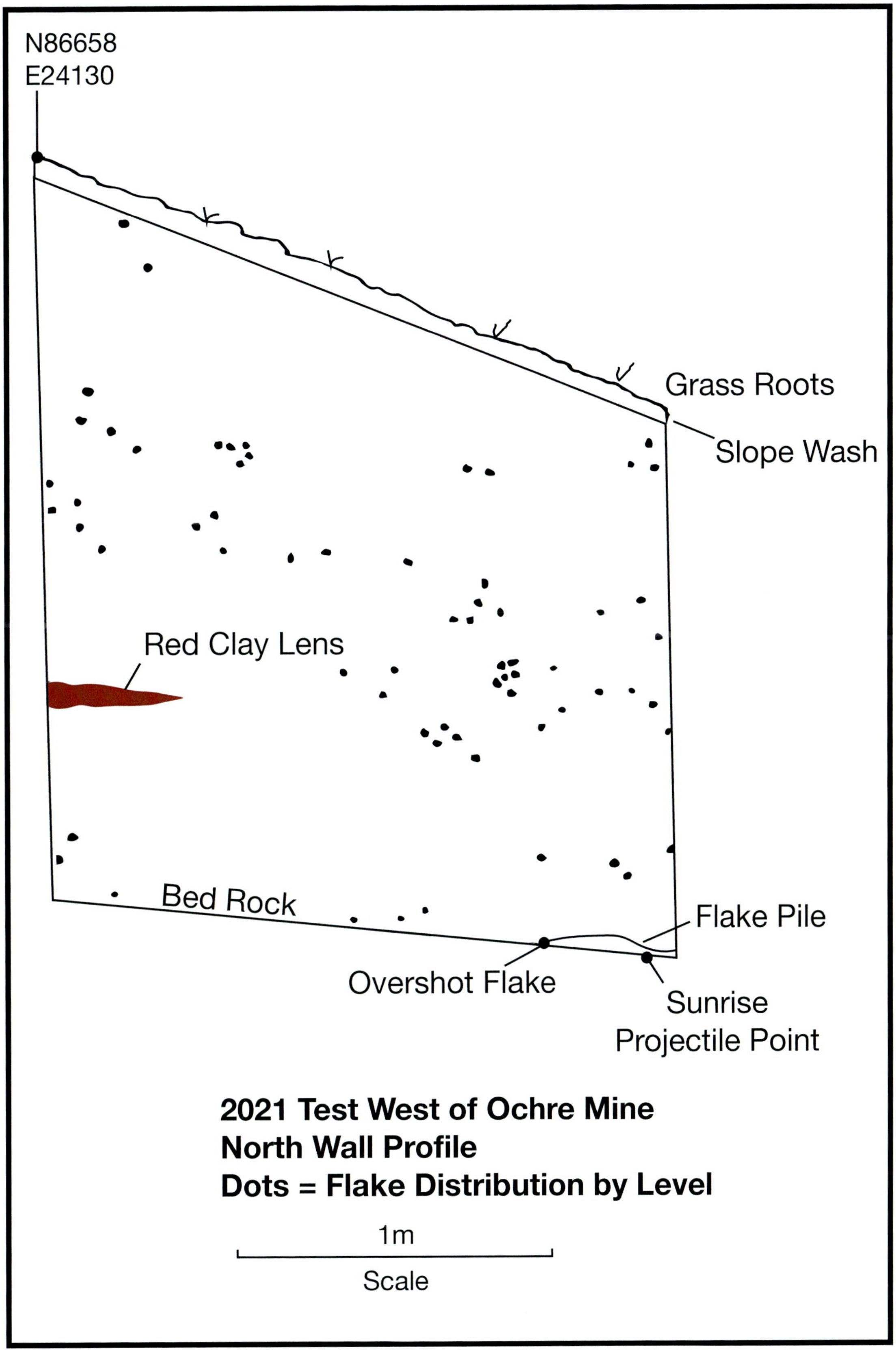

Figure 5-8. Profile 2 from the OCM-21-1 test west of the trench feature

Chapter 6

Red Ochre, Chert, and Quartzite at the Sunrise Iron Mine / Powars II Site

Wayne M. Sutherland

Introduction

The unique and significant occurrences of red ochre and chert/quartzite at the Sunrise Iron Mine / Powers II Site prompted George Frison to request my investigation into the character and origins of these deposits. Frison described the chert/quartzite nodules as some of the finest stone tool material found anywhere in the world and the red ochre as totally unique. I had already studied the iron mine during investigations of iron resources and rare earth elements while working at the Wyoming State Geological Survey. Several visits to the sites with George, coupled with his discussions of the materials and his enthusiasm, further opened my curiosity and led to this investigation.

Hartville Uplift and Sunrise Iron Mine

The Hartville Uplift is an elongated north-northeast trending dome approximately 64 km (40 mi) long. A high angle fault forms the boundary for the east side; the Hartville Uplift is the western, up-thrown block. Rock outcrops in the uplift include the Precambrian Whalen Group, overlain by the Mississippian-Devonian Guernsey Formation, which is overlain by the Pennsylvanian Hartville Formation (Figs. 6-1., 6-2.). Copper ore was mined from the Sunrise Mine from 1880 to 1887, and iron was mined from 1898 to 1980 (Sutherland and Cola, 2015).

Iron deposits are primarily hosted by the Whalen Group, a thick sequence of green schists, quartzites, dolomites, micaceous schists, and graphitc schists (Wilson, 1976). The lower part of the Good Fortune Schist, comprising impure hematite and limonite-stained flint, encloses lenticular bodies of red iron ore immediately above the uppermost limestone of the older Precambrian rocks (Frey, 1947). A thin layer of iron-stained, siliceous schist separates the iron ore from the limestone. The Good Fortune Schist is faulted and is part of a steeply east-plunging synform. Ore bodies appear localized in cross-folded zones superimposed on the synform in the thickest part of the Good Fortune Schist. Contacts between low-grade hematite and the wall rock are both sharp and gradational.

Frey (1947) interpreted the iron ore as a replacement of the schist. Relatively soft, light-red, fine-grained, generally schistose iron is commonly referred to as "paint ore." Fine-grained, hard, dark bluish-gray, compact ore with a smooth fracture is referred to as "blue ore." The paint ore is deep red, somewhat greasy to the touch, and leaves a rusty, metallic stain on contact. Paint ore averages 57 percent Fe_2O_3 while the schist below it contains only 8.5 percent Fe_2O_3. Blue ore is a dense, bluish-black hematite with a rusty surface and yields about 89 percent Fe_2O_3 (Sutherland and Cola, 2015). Sims and others (1997) indicate that most of the iron ore is derived from hydrothermal alteration of

Figure 6-1. Sunrise Area Geologic Units
(After Harris and others, 2005)

Qsl — **Silty loess (Holocene/Pleistocene)**—Brown to dark brown deposits of windblown silt with minor clay and very fine sand; includes reworked deposits of loess. Thickness may exceed 9 m (30 feet) in valley deposits in and tributary to Pleasant Valley, Webb Canyon, and Whalen Canyon. Locally called "blow dirt"

Qg — **Gravel deposits (Pleistocene)**—Boulder to pebble conglomerate deposited in a fluvial environment by the ancestral North Platte River. Contains fluvial cross-bedded sand zones. The boulders include many types of crystalline, metamorphic, and sedimentary rocks including but not limited to granite, monzonite, amphibolite, schist, quartzite, marble, limestone, and sandstone. One fragment of black jade measuring 5 cm x 5 cm x 7.5 cm (2" x 2" x 3") was found in a gravel pit east of Guernsey. Includes clasts from the Hartville uplift as well as clasts from the Laramie Mountains and farther west. Found mostly in the southern half of the map area, concentrated near north-south trending faults. The gravel deposits attain local thicknesses in excess of 46 m (150 feet)

Ta — **Arikaree Formation (Miocene)**—Light gray tuffaceous sandstone containing occasional limy concretions; vertebrate fossils common in some areas; and forms bluffs and narrow canyons with local areas of badland topography. Only the upper part of the Arikaree Formation (McGrew, 1963) occurs in the area

$\mathbb{P}h^{2,3}$ — **Hartville Formation, divisions 2 and 3 (Pennslyvanian)**—Interbedded gray limestone, buff to chalky white limestone and dolomite, pink dolomite, buff Aeolian (Tom Ahlbrandt, U.S. Geological Survey, personal communication, 1995) gray, red, and maroon silt and claystones, and thin black shales. Divisions 2 and 3 are combined and mapped together following Denson and Botinelly (1949) because the dolomites in division 2 grade laterally to massive limestones like those in division 3. Brachiopods are common in limestone and dolomite layers and fusulinids are reported in fine-grained clastic layers (Denson and Botinelly, 1949). Forms ledged slopes and cliffs commonly covered with Mountain Mahogany. About 90 m (300 feet) thick in the area.

$\mathbb{P}h^{4,5}$ — **Hartville Formation, divisions 4 and 5 (Pennslyvanian)**—Interbedded maroon, red, pink, and gray siltstones and claystones; gray, brown, and buff limestone, pink dolomite, and thin gray sandstones. Divisions 4 and 5 are combined and mapped together following Denson and Botinelly (1949). Forms smooth slopes with limestone outcrops; about 75 m (250 feet) thick in the area

$\mathbb{P}Mh^{6}$ — **Hartville Formation, division 6 (Pennslyvanian/Mississippian)**—Well-indurated maroon to red to locally white orthoquartzite; forms cliffs and rocky knolls; deposited on a well-developed karst surface, and fills sinkholes and caverns in the underlying Guernsey Formation; thickness 0 to 37 m (0 to 120 feet)

$\mathbb{P}Mh^{4,5,6}$ — **Hartville Formation, divisions 4, 5, and 6 (Pennslyvanian/Mississippian)**—Mapped together in the Haystack Hills where the Hartville division 6 occurs as a thin red band or is locally not present

MDg — **Guernsey Formation (Mississippian/Devonian)**—Gray limestone with local beds and zones of chert; brachiopods common; thickness 0 to 55 m (0 to 180 feet). Forms smooth to ledged slopes commonly covered with Mountain Mahogany. Sando and Sandberg (1987) considered all except the lowermost part of this unit equivalent to members at the Madison Limestone and rejected the use of Guernsey Formation as being superfluous

Єq — **Quartzite of Devonian or Cambrian age**—Gray to red to cream-colored, coarse-grained cross-bedded orthoquartzite found in lenses locally throughout the area; possibly equivalent to the Deadwood Quartzite (Cambrian) or the lowermost part of the Englewood Formation and the Fremont Canyon Sandstone (Devonian), based on work by Sando and Sandberg (1987)

Whalen Group (Early Proterozoic or late Archean)—As used by Sims and others 1997, whose descriptions of rock units are used below

XWm — **Metabasalt**—Dark-green, fine-grained actinolite-biotite-chlorite schist; flows are locally pillowed. Estimated thickness 100 to 150 m (330 to 490 feet)

XWd — **Metadolomite**—Gray, pink, white, and yellow, medium-grained dolomite and marble, tremolite dolomite, siliceous dolomite, and local limestone. Tremolite occurs as radiating blades resembling turkey tracks. Contains abundant concentrically layered algal stromatolitic mounds. Estimated thickness 300-500 m 1,000 to 1,600 feet)

XWdp — **Metapelite**—Thin, gray to brown, plagioclase-quartz-biotite-muscovite schist, interpreted as pelite; occurs near base of dolomite unit (XWd) in Graves Ranch anticline

XWf — **Ferruginous schist**—Includes biotite-chlorite schist, phyllite, and ferruginous quartzite

XWfh — **Hematite bodies**—Hematite bodies at top of ferruginous schist unit (XWf) were mined at the Good Fortune, Sunrise, and Chicago mines and consist of specularite and associated colloform hematite (Ebbett, 1956)

XWfi — **Banded iron formation**—of Lake Superior type

XWb — **Quartzofeldspathic schist**—Primarily east of Wheatland fault system; contains garnet, sillimanite, and rare andaluste and pyrite. Graded beds common

Figure 6-1. Sunrise Area Geologic Units

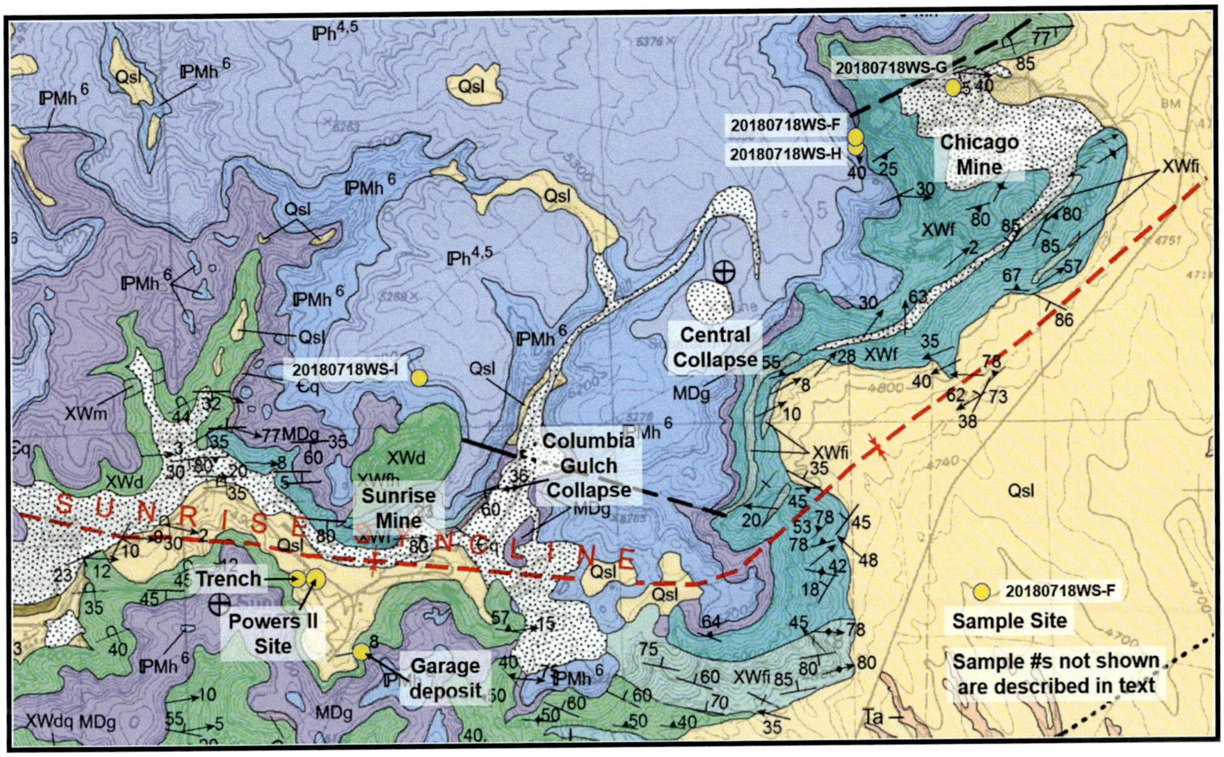

Figure 6-2. Sunrise Geology, Locations and Samples.

banded iron formation (BIF).

Although the major ore bodies occur within Precambrian rocks, iron is also found within the overlying Paleozoic rocks. Sims and others (1997) suggested that some of the hematite may be due to leaching from the overlying limestones in the Guernsey Formation during carbonate dissolution by meteoric water. Hematite fills solution cavities and brecciated fractures extending upward into the overlying Mississippian-Devonian Guernsey Formation and the Pennsylvanian-Mississippian Hartville Formation. The Guernsey Formation is dominated by gray limestone, and the lower part of the Hartville Formation is dominated by fine-grained sandstone and quartzite accompanied by minor limestone layers (Harris and others, 2005). Iron contacts are relatively sharp within the Guernsey limestone, but iron enrichment penetrates sandstone adjacent to some fractures within the Hartville Formation. Outcrops in the Columbia Gulch collapse, the large Central collapse, and in the wall cut north of the Chicago mine face demonstrate these iron relationships within the Paleozoic rocks.

Red Ochre

Solid, fine-grained, soft red hematite ochre pieces associated with the Powars II site are interpreted to have been prehistorically mined for use as paint or dye (Frey, 1947; Miller, 2009; George Frison, personal communication, 2014; Frison et al. 2018). These pieces of ochre may be hematite from partings in the Good Fortune Schist, deposited as described by Sims and others (1997). They also could result from fine material concentrated by more recent surficial water, similar to the tailings pond fines from the old Sunrise beneficiation plant. Anecdotal evidence suggests that some caves in the area were partially filled with red ochre, which would have been deposited by meteoric water. Iron oxide concentrated by surficial water moving down the Chicago highwall also appears similar to the red ochre pieces from the Powars II site. No samples were taken directly from partings in the schist, however, samples taken from immediately below the Powars II site were evaluated using x-ray fluorescence (XRF). These include red ochre mixed with clay, and water settled fines extracted in the laboratory from red ochre-mud. Attached spreadsheets show analyses from both XRF and outside chemical analyses.

Similar red ochre occurs north of Rawlins at a site known as the Rawlins Red iron deposit (Sutherland and Cola, 2015). Botryoidal and stalactitic hematite found at Rawlins Red indicate deposition by water movement along a fault. The suggested source for this iron deposit is re-mobilization from the Amsden Formation and local precipitation into the Madison Limestone (partial Guernsey Formation equivalent) and slivers of the Flathead Sandstone (Deadwood Quartzite equivalent). Sutherland and Cola (2015) suggest that Rawlins Red might have been of interest to Paleoindians for use as a pigment. However, extensive surface disturbances since mining began in 1870 probably obscured any indications of prehistoric use. Two sample analyses of Rawlins Red hematite are provided for comparison.

Red ochre pieces, provided by George Frison and associated with archaeological materials (from a trench above the Powars II site near the stone rings) have a slightly different character than water-laid ochre from the Chicago highwall, the tailings ponds, or from immediately below the Powars II site. These pieces are more coherent, do not crumble easily, and have a character similar to a soft crayon. This crayon-like quality was puzzling as to whether it was the result of a natural or artificial binding material.

The same material was found in-place in July 2019 in a massive form in a trench cut beneath mine tailings. This material exhibits a weak conchoidal to micaceous fracture in some masses. X-ray diffraction of the massive ochre shows that it is composed mostly of hematite, accompanied by a small amount of illite clay. The small amount of illite appears to form a natural binder to give the ochre its cohesive quality not found in simple water-deposited fine iron oxide. This massive Powars II ochre essentially contains no other minerals. The lack of other minerals, such as quartz, also distinguishes it from similar-appearing iron oxides that do not have the cohesive character of the Powars II site ochre.

Cherts and Quartzites

Silicification, including cherts and quartzites are relatively common across the Hartville Uplift within rock units of varying ages. Precambrian occurrences include discontinuous quartzites within dolomite and sandy dolomite overlying mica schist, microcrystalline ferruginous quartzite associated with BIF, and multicolored chert and jasper veins associated with fault and fracture zones (Millgate, 1965). In addition, a low quality, opaque, light-green opal occurs within at least one of the major iron bodies and is observable in the north wall of the Sunrise glory hole. Hydrothermal alteration of these various silica-rich bodies could silicify parts of the overlying rock units. Harris and others (2005) describe lenses of Devonian-Cambrian quartzite, possibly equivalent to the Cambrian Deadwood Quartzite or silicified parts of the Flathead Sandstone. A 2018 visit to the area identified cross-bedded quartzite with an appearance similar to silicified conglomerate observed within the Flathead Sandstone in several other parts of Wyoming.

The Mississippian-Devonian Guernsey Formation hosts chalcedony (agate) fracture fills and silica speleothems in caverns within limestone layers. Sandstones in the Hartville Formation are locally silicified and vary from silica-cemented sandstones to true quartzites and some nodular cherts. The Miocene Arikaree Formation, dominated by tuffaceous sandstones, overlies the Hartville Formation (Harris and others, 2005) and hosts agates in some locations. The Arikaree covered the entire Sunrise area during the Miocene, but has since been eroded, leaving only scattered remnants. Miocene and Oligocene tuffaceous sandstones are known hosts for chert and agates in many parts of Wyoming (Hausel and Sutherland, 2000). Tuffaceous sandstones, such as in the Arikaree Formation, are potential sources for silicification in older, underlying rock units.

Chert nodules, which appear to be the same material used to fashion Clovis-age, Sunrise artifacts, were exposed in a backhoe trench a few hundred feet west of the Powars II archaeological site in the valley bottom. The nodules are blue to cream and multi-colored chert/chalcedony with opaque white siliceous coatings. The nodules, often of irregular, but somewhat rounded shapes, vary in size up to several feet across. The morphology of some boulders includes very narrow blade-like extensions beyond the rounded masses, which suggests that these formed in-place. Any tumbling, such as in streams, would have likely removed these irregularities.

The matrix surrounding some of the irregular nodules is a soft, red to brown, iron-rich clay, which appears to be in-place residual material. Other areas of the matrix include angular sand- and gravel-sized particles. The overall matrix was mapped as silty loess by Harris and others (2005). The sand- and gravel-sized particles include: chips and pieces of chert nodules, jasper, hard hematite masses, soft hematite of the same character as ochre nodules found at the Powars II site, hematitic and limonitic masses with particles of broken chalcedony, and rare pieces of quartz.

The coarse part of the matrix appears to be mechanically emplaced in intimate and poorly defined contact with the residual clays. George Zeimens (personal communication, 2021) noted that many of the chert/quartzite boulders were pushed into the bottom of the draw during modern iron mine activity. This suggests that more of the valley bottom, prior to iron mine activity, may have contained exposed cobbles and boulders that could have been easily used for stone tools in the Powars II site. The valley bottom probably contained a layer of boulders in clay, of which only a remnant remains in the bottom of the draw. Erosion of the clay across the valley would have left residual boulders and cobbles that were easily accessible for Clovis-age use. Historical construction activity then pushed remaining boulders into the draw and out of the way for historic mine-related developments.

The clay and chert/quartzite boulder deposit in the valley bottom suggests that it relates to silica-rich meteoric water descending from overlying tuffaceous sediments. These sediments include Tertiary formations, such as the Oligocene White River Formation and the Miocene Ogallala Formation, that covered the area prior to their removal by erosion. An identical chert was found in a similar, yellow rather than red, clay host north of the Chicago mine (George Frison, personal communication, 2016).

Frison also noted that unique pale-blue to multicolored chert nodules, similar to those at Sunrise, are found below Spanish Point in the Bighorn Mountains near Johnny Creek. Those cherts occur within a fine-grained soft matrix similar to that found at Sunrise, but with significantly less iron content. The common factor between these two distant locations is the previously overlying Miocene or Oligocene tuffaceous sediments, which could provide silica through percolation of meteoric waters. No chert samples from Spanish Point were tested, but trace element differences would be expected due to differences in adjacent rock units and related elements potentially contributed through water movement.

Detailed examination of the microstructure of the chert nodules has not been investigated. However, microscopic structural details are hypothesized to be significant factors in the characteristics of the material for use in high-quality stone tools. Identification of the microcrystalline to nanocrystalline polymorphs of SiO2 present, such as tridymite, cristobalite, moganite, and others, may serve to identify material specific to the Sunrise/Powars II site.

Localized quartzite and minor limestone layers occur within parts of the Hartville Formation (Harris and others, 2005). A dark-gray chalcedony in a blue to purple and green quartzite lens within thin limestone crops out about 0.23 km (0.15 mi) north of the Sunrise glory hole. This material is a source for some stone tools found at the Powars II site (George Zeimens, personal communication, 2018). The high-quality, compact quartzite is of limited extent in a shallow draw and retains some of the character typical of other quartzites within the Hartville Formation. The site is underlain by fine-grained, finely layered silicified sandstone and quartzite in the lower part of the Hartville Formation. The position of the site in a draw suggests that groundwater movement related to a fracture may have contributed to enhanced silicification at that location.

Chemical Analyses

Geochemical analyses, including whole rock chemistry and trace elements, were completed by ALS Chemex of Reno, Nevada for samples submitted by the Wyoming State Geological Survey (WSGS). Geochemical analyses on samples included whole rock analyses (major element concentrations in the form of oxides) by inductively coupled plasma, atomic emission spectrometry or mass spectrometry, and atomic absorption. These samples were collected and analyzed in conjunction with statewide surveys of potential sources of rare earth elements (Sutherland and others, 2013; Sutherland and Cola, 2016) and iron resources (Sutherland and Cola, 2015).

More recently, samples were analyzed by the WSGS with in-house X-ray fluorescence using an Olympus Vanta M series handheld X-ray fluorometer (XRF). XRF results generally include multiple measurements of either heterogeneous specimens or powdered samples. Concentrations listed are averages in parts per million (ppm) from multiple measurements in several circular focus areas about 4.5 mm across, which may not be representative of the entire sample. Light elements (H through Na in the periodic table) are not evaluated with XRF.

Silver (Ag) is noted in two ochre nodules from the Powars II site and is present in some (but not all) iron oxide and related red clays from below the Powars II site and in matrix from the backhoe trench. It is also elevated in some (not all) chert and jasper from the backhoe trench. As, Cr, Cu, Sn, V, and Zn appear to follow similar patterns. None of these elements individually characterizes the site or the trench.

Additional samples of jasper and quartzite from McCann Pass, about 6.4 km (4 mi) northeast of the Chicago Mine are included in the analyses to provide a regional comparison. However, these show little chemical commonality with chert and jasper from the Sunrise area.

Statistical analyses of combinations of these elements might demonstrate a preferred chemical profile useful in comparison with material from other locations. Iron oxide, concentrated by surficial water on the Chicago mine face shows a similar profile to iron oxides from Sunrise and material from the backhoe trench. However neither chert nor jasper from the Chicago mine, nor low quality opal from the Sunrise mine, nor quartzite/chert from the Guernsey Formation north of the Sunrise glory hole appear to follow the same pattern. Iron oxide fines and clay retrieved from the Sunrise mine settling pond and iron oxide from the Rawlins Red iron deposit (similar in appearance to the ochre from Powars II site) do not contain Ag, nor do they host a similar combination of trace elements, nor do they contain illite clay.

These trace element analyses of ochre, related materials, chalcedony, and jasper provide only indications of chemical patterns based on a very small number of samples. Before diagnostic chemical patterns can be demonstrated for materials from the Powars II site, a large number of samples must be evaluated.

See Tables.

Summary and Conclusions

The complex interrelationships of hematite and silicification in the Sunrise area may derive from multiple processes. Chert and jasper veins associated with fault and fracture zones and low quality opal within the major Sunrise iron deposit likely are products of hydrothermal activity. This is the same process that modified BIF to more massive iron (Sims and others, 1997). A secondary hematite source from leaching of overlying limestones during carbonate dissolution in the Guernsey Formation by meteoric water (Sims and others, 1997) may have contributed to the formation of red ochre. However, this does not explain the extension of iron deposition upward into fractures in the overlying Hartville Formation, which may represent yet another phase of water movement.

Red ochre nodules, with the character of very soft crayons, found at the Powers II site, originate either from upward movement of iron-oxide in hydrothermal waters in one or more pulses, or in deposition from descending meteoric waters. Illite clay was found to be the binder, which makes the red ochre nodules unique. However, the source and genesis of the illite clay was not determined.

The morphology of the large chert/quartzite boulders in the trench near the Powars II site indicates that some of them formed in-place. Surrounding matrix for other boulders includes both clay and angular non-clay materials. This, combined with historical evidence indicates that some boulders were pushed into the bottom of the draw during modern iron mine activity. These boulders and cobbles probably formed similarly in the valley bottom and were subsequently exposed by erosion of the surrounding clay. Such exposed boulders and cobbles would have been a readily available source for stone tool material. Later historical activity pushed remaining boulders into the draw to accommodate mine-related developments.

A similar chert occurrence in the Spanish Point area in the Bighorn Mountains, combined with other tuffaceous sediment-related chert occurrences across the state (Hausel and Sutherland, 2000) suggests a common derivation. The sunrise chert nodules most likely formed from silica-rich meteoric water descending from overlying tuffaceous sediments.

Trace elements such as Ag and others within the ochre and chert nodules apparently derived from local sources. Trace element analyses provide a starting place for chemical characterization of these materials. However, insufficient numbers of analyses were available to fully characterize trace element patterns and apply them to materials from the Powars II site.

2019-7-25 Sunrise Chemistry p. 1

2019 XRF & previous outside analyses

						Whole Rock Analyses			
	[ALS = outside chemical analysis; XRF = WSGS analysis]	XRF or				SiO2	Al2O3	Fe2O3	CaO
Sample ID	**Notes**	**ALS**	**Date**	**Lat.**	**Long.**	%	%	%	%
	Ochre & related FE-Oxides								
Sunrise-01	G. Ziemens' ochre nodule 1 from Powers II archaeological site	XRF	1/14/2019	42.3318	-104.7077				
Sunrise-02	G. Ziemens' ochre nodule 2 from Powers II archaeological site	XRF	1/14/2019	42.3318	-104.7077				
20190725GF-1	Sample of massive ochre from trench cut through tailings on bench NW of 2017 dig (approx. location)	XRF	7/25/2019	42.3317	-104.7074				
20180718WS-J	tailings pond fines (Sunrise-SP)	XRF	1/15/2019	42.3316	-104.7040				
20180718WS-F	red ochre on on the Chicago highwall concentrated by rainwater moving down along the cut face	XRF	2019	42.3462	-104.6829				
20140624WS-B	red ochre/mud below Powers II archaeological site (settled fines)	XRF	2019						
20140926WS-C	red ochre, iron oxide plus clay, from below Powers II archaeological site	ALS	2014	42.3318	-104.7077	16.95	6.99	57	6.96
20150622LC-1B	red, iron-rich matrix clay from backhoe trench surrounding chert with some jasper	ALS	2015	42.3315	-104.7080	42.7	1.06	47.2	0.32
20180718WS-A	soft, fine red clay material from adjacent to chalcedony cobbles & boulders in backhoe trench	XRF	2019	42.3315	-104.7080				
20120322WS-4	Red, ultra-fine, settling pond tailings	ALS	2013	42.3316	-104.7040	12	5.2	75.2	1.52
20120322WS-3	1/4" minus settling pond tailings plus clay	ALS	2013	42.3316	-104.7040	42.3	13.6	31.7	2.04
CC136B	Rawlins Red soft, massive, botryoidal to stalactitic hematite with minor limonite.	ALS	2014	42.8528	-110.6604	1.62	0.48	89.2	3.82
CC135	Rawlins Red Fe-oxide-rich, fine-grained sandstone, shale & minor fine-grained arkosic conglomerate	ALS	2014	44.7174	-109.3585	3.87	0.3	90.5	0.35
CC134	Rawlins Red Fe-oxide-rich, fine-grained sandstone, shale & minor fine-grained arkosic conglomerate	ALS	2014	44.7186	-109.3596	57.1	5.17	30.8	0.73
	Opal								
20120322WS-1	pale, light-green opal from tailings pile; large body of same apparent in pit highwall	ALS	2013	42.3316	-104.7040	44.1	31.4	6.42	0.73
	Chalcedony, chert, jasper								
20150622LC-1A	sharp, irregular shaped blue to cream & multi-colored chert/chalcedony nodule from backhoe trench	ALS	2015	42.3315	-104.7080	98.6	0.18	0.73	0.08
20150622LC-1Aa	cream to light-blue chert nodule from backhoe trench	XRF	2019						
20150622LC-1Ab	gray to mottled pink & red-brown chert nodule from backhoe trench	XRF	2019						
20180718WS-B2	opaque white coating on chert nodules from backhoe trench	XRF	2019						
20180718WS-H	Chicago Mine pink to greenish-gray & blue, mottled chalcedony (similar to that in trench below the Powers II site)	XRF	1/14/2019	42.3462	-104.6828				
20180718WS-G	cut wall N. of the Chicago face: reddish-pink, brown, & minor cream jasper & chalcedony from within Fe-rich breccias	XRF	1/14/2019	42.3469	-104.6792				
20180718WS-B	gray to black & light-brown chalcedony from backhoe trench	XRF	1/14/2019	42.3315	-104.7080				
20120322WS-5a	limonite and orange-brown jasper; hydrothermal deposit on road to Chicago mine	ALS	2013	42.3404	-104.6790	90.1	0.39	4.76	1.16
20120322WS-5b	hematite plus black jasper and calcite; hydrothermal deposit on road to Chicago mine	ALS	2013	42.3404	-104.6790	85.5	0.35	10.6	0.97
20180718WS-C	reddish-brown to yellowish-brown jasper from backhoe trench	XRF	1/14/2019	42.3315	-104.7080				
20150622LC-1C	iron-rich jasper & quartzite from backhoe trench	ALS	2015	42.3315	-104.7080	47.2	0.38	48.4	0.1
	Hartville Formation								
20180718WS-I	Hartville Fm; dark-gray chalcedony in blue to purple & green quartzite layer within carbonate N. of Sunrise glory hole	XRF	1/14/2019	42.3384	-104.7028				
	Guernsey Formation								
20180718WS-E	Guernsey Fm basal arkose; x-bedded, silicified, Fe-rich jasper & qtz conglomerate from the Chicago mine area	XRF	1/15/2019						
	Whalen group hematite bodies (iron ore)								
20180718WS-D	dark blue-gray to reddish-brown iron ore; minor white chalcedony blebs from within breccia at Powers II site	XRF	1/14/2019	42.3318	-104.7077				
20140926WS-A	garage deposit, ~1,260 ft SSE of Powers II site; metallic, hematite-rich, non-magnetic rock with some red iron oxide	ALS	2014	42.3283	-104.7062	8.6	0.19	89.1	0.56
20120322WS-6	Chicago mine highwall iron ore breccia	ALS	2013	42.3451	-104.6810	58	13.7	19.35	0.29
	Whalen Group Silver Springs (Good Fortune) Schist (ferruginous)								
20140624WS-C	iron schist under red ochre source slope	ALS	2014	42.3318	-104.7077	65.7	13.55	8.46	2.15
20140926WS-B	Chicago Mine Fe-rich schist, includes variation of oxides (orange-red), metallic minerals, brecciated	ALS	2014	42.3318	-104.7077	40.2	7.97	39.5	2.64
	Jasper from McCann Pass NE gossan ~ 4 mi NE of Chicago face								
20150923LC-2	McCann Pass NE gossan ~ 4 mi NE of Chicago Mine; jasperized rock with metallic coating & botryoidal iron (goethite)	ALS	2015	42.3771	-104.6227	72.3	0.61	18.35	0.09
20150923LC-3	McCann Pass NE gossan ~ 4 mi NE of Chicago Mine; rustly-limonitic-metallic quartzite, rusty/metallic veins, tan-yellow shale	ALS	2015	42.3771	-104.6227	60.6	13	7.42	0.32

2019-7-25 Sunrise Chemistry p. 2

2019 XRF & previous outside analyses

Sample ID	MgO	Na2O	K2O	Cr2O3	TiO2	MnO	P2O5	SrO	BaO	C	S	LOI	Total	Ag	Al	As	Ba	Bi	Ca	Cd	Ce	Co	Cr	Cs
	%	%	%	%	%	%	%	%	%	%	%	%	%	ppm	%	ppm	ppm	ppm	%	ppm	ppm	ppm	ppm	ppm
Sunrise-O1														5.42	3.74	65.38	3.72	63.74	0.35	0	0	0	2,062	
Sunrise-O2														2.84	3.41	63.96	0	84.18	0.30	0	0	0	2,159	
20190725GF-1														1.8	3.25	49.2	63.3	65.2	0.39	0	0	0	2,393	85.5
20180718WS-J														0	1.71	98.9	52.8	43.18	0.74	0	0	0	1,585	
20180718WS-F														29.5	0.43	150.5	13.2	77.2	0.28	4.9	43.6	0	3,052	
20140624WS-B														13.2	2.04	67.9	331	28.8	0.94	2.3	44.8	197	3,048	
20140926WS-C	0.67	0.06	2.26	0.01	0.24	0.13	0.15	0.04	0.05	1.79	0.02	6.89	98.4	0		68.2	470	0.07		1.9	20.2	104	70	3.85
20150622LC-1B	0.59	0.04	0.08	<0.01	0.2	0.04	0.25	0.08	0.14	0.04	0.14	6.9	99.6	<0.5		148	1,220	0.05		<0.5	22.3	15	30	0.51
20180718WS-A														19.2	0.73	263	230	75.1	0.33	3.6	56.2	0	2,787	
20120322WS-4	0.49	0.18	1.57	0.03	0.12	0.08	0.17	0.04	0.03	0.44	0.05	3.66	100.29	0		58.6	209	0.05		1	16.6	60	30	2.77
20120322WS-3	0.91	0.04	4.79	0.01	0.2	0.06	0.13	0.12	0.03	0.64	0.03	4.85	100.78	0		54.7	264	0.07		0.5	21.4	42	70	7.1
CC136B	0.07	0.02	0.06	0.04	0.01	0.08	0.03	<0.01	<0.01	0.83	<0.01	3.41	98.84	<0.5		6.1	36.3	0.3		2.4	29.8	8	290	0.45
CC135	0.08	0.02	0.11	0.02	0.01	<0.01	0.08	<0.01	0.01	0.09	<0.01	0.56	95.91	<0.5		12.8	53.2	0.02		0.8	6.7	2	130	0.19
CC134	0.2	0.05	1.63	0.01	0.16	0.02	1.47	0.72	0.04	0.04	0.13	2.13	100.23	<0.5		6.5	364	0.38		0.5	946	6	80	1.83
20120322WS-1	0.43	0.03	9.42	0.01	0.02	0.07	0.2	0.22	0.03	0.19	0.05	6.61	99.69	0		22.9	284	0.02		0.7	14.3	60	30	15.2
20150622LC-1A	0.03	0.04	0.01	0	0.05	0.01	0.01	0.01	0.04	0.02	0.01	1.6	101.39			1.6	337	0.01		0	0.9	1	30	0.09
20150622LC-1Aa														2.9	0	10.1	26.3	11.6	0	2.7	75	0	2,845	
20150622LC-1Ab														2.3	0	3.1	788	8.7	0	5.1	43.9	0	2,618	
20180718WS-B2														3.8	0.10	16.2	1,163	31.4	0	2.6	117	0	3,318	
20180718WS-H														0	0	0	0	0	0	0	0	0	0	
20180718WS-G														0	0.20	2.43	462.47	0	29.07	0	0	0	0	
20180718WS-B														0	0	0	431.36	0	0	0	0	0	0	
20120322WS-5a	0.18	0.03	0.09	0.01	1.18	0.03	0	0	0.01	0.52	0	2.74	100.68	1.2		20.8	68.8	0.33		0.5	4	3	50	1.65
20120322WS-5b	0.3	0.02	0.09	0.01	0.17	0.01	0.08	0.01	0.01	0.22	0.02	3.1	101.22	0		67.3	117.5	0.06		0.6	8.2	0	80	1.57
20180718WS-C														3.76	0.06	190.36	310.1	0	0.02	0	0	0	725.9	
20150622LC-1C	0.04	0.03	0.02	<0.01	0.01	0.06	0.1	0.03	0.03	0.02	0.09	3.6	100	2.2		84.2	298	0.01		<0.5	5	22	10	0.28
20180718WS-I														0	0	0	150.8	0	0	0	0	0	0	
20180718WS-E														0	4.16	76.32	52	20.96	0.14	0	23.3	0	1,133	
20180718WS-D														0	0.58	63.52	0	75.22	0.39	0	0	0	1,880	
20140926WS-A	0.03	0.01	0.01	0	0.01	0.01	0.15	0.07	0	0.17	0.02	0.5	99.24	0		31.2	25.8	0.02		1.9	1.7	5	10	0.05
20120322WS-6	0.68	0.12	5.78	0.02	0.52	0.03	0.05	0.02	0.03	0.13	0	2.47	101.06	0		41.6	276	0.27		0	78.4	28	140	4.67
20140624WS-C	0.93	0.05	6.35	0.03	0.58	0.02	0.13	0.01	0.05	0.47	0.01	2.69	100.7	0		28.4	490	0.07		0	30.8	25	200	2.35
20140926WS-B	2.29	0.08	1.95	0.01	0.32	0.04	0.17	0.01	0.02	0.97	0	5.1	100.3	0		29.9	157.5	0.37		1.1	30.2	28	70	7.05
20150923LC-2	0.11	0.02	0.18	<0.01	0.25	0.02	0.05	<0.01	0.01	3.99	0.11	7.1	99.09	1		47.8	59.1	5.18		0.5	7.7	<1	30	0.26
20150923LC-3	0.27	1.3	7.09	0.01	0.3	0.02	0.07	0.02	0.1	4.44	0.25	7.87	98.39	0.7		23.9	882	0.47		1.4	68.9	3	50	3.38

2019-7-25 Sunrise Chemistry p. 3

2019 XRF & previous outside analy Blue numbers for % Fe are calculated from analyses showing Fe2O3 [Hematite Fe2O3 _ 69.94 % Fe]

	Cu	Dy	Er	Eu	Fe	Ga	Gd	Hf	Hg	Ho	In	K	La	Li	Lu	Mg	Mn	Mo	Nb	Nd	Ni	P
Sample ID	ppm	ppm	ppm	ppm	%	ppm	ppm	ppm	ppm	ppm	ppm	ppm	ppm	ppm	ppm	%	ppm	ppm	ppm	ppm	ppm	ppm
Sunrise-O1	311				59.77				0			6,790	0			2.28	5,176	7.66	2.88	0	0	498
Sunrise-O2	401				64.38				0			5,532	0			2.97	2,787	7.96	5.2	0	0	619
20190725GF-1	366				65.30				0			4,983	0			2.61	3,332	10.7	7.6	0	0	447
20180718WS-J	380				45.52				0			12,283	0			1.00	1,933	0	0	0	0	2,886
20180718WS-F	658				56.64				0			510	14.8			0.47	5,010	12.9	7.7	29.0	3,199	1,919
20140624WS-B	376				19.47				32.4			13,970	0			0.24	3,802	10.9	39.1	68.1	2,893	1,736
20140926WS-C	19	3.13	1.73	0.87	**39.87**	8.1	3.81	1	0.036	0.62			10.1	10	0.24			5	14.5	11.8	501	
20150622LC-1B	1,930	3.78	2.23	1.53	**33.01**	2.9	6.3	0.3	0.434	0.77	0.039		12	<10	0.33			12	11.2	15.1	127	
20180718WS-A	2,772				26.98				16.1			1,585	29.5			0.47	4,445	21.8	26.3	41.0	2,608	3,218
20120322WS-4	18	3.4	1.86	0.94	**52.59**	3.9	4.34	1	0.078	0.67			9.2	0	0.28			1	4.3	9.4	454	
20120322WS-3	8	3.04	1.63	1.16	**22.17**	6.4	4.54	1.4	0.164	0.59			11.5	0	0.26			0	3.5	13.6	263	
CC136B	20	1.79	0.64	0.58	**62.39**	3.5	2.71	<0.2	<0.005	0.29			2.6	10	0.05			56	<0.2	7.8	36	
CC135	<1	0.76	0.43	0.19	**63.30**	5.6	0.89	0.3	<0.005	0.17			1.2	10	0.05			109	<0.2	2.4	96	
CC134	6	4.78	1.84	4.91	**21.54**	8.6	15.2	3.3	<0.005	0.72			337	10	0.25			4	4.6	209	21	
20120322WS-1	7	1.62	0.56	1.59	**4.49**	2.1	5.4	0.7	0.041	0.23			6.4	0	0.12			0	0	11.7	296	
20150622LC-1A	16	0	0.03	0.04	**0.51**	0.7	0.13	0	0.118	0.02	0		0.5	0	0.01			3	0.3	0.6	3	
20150622LC-1Aa	66				0.18				27.6			0	79.6			0	3,200	2.4	0.6	425	1,055	0
20150622LC-1Ab	78				0.27				20.0			0	63.1			0	3,748	0.6	0.5	65	1,177	0
20180718WS-B2	88				0.36				29.7			0	14.9			0.04	3,135	2.4	4.5	137	889	306
20180718WS-H	0				0.05				0			0	0			0	0	0	0	0	0	0
20180718WS-G	0				1.23				0			510.0	0			0.08	560	0	3.72	0	91.77	44.23
20180718WS-B	0				0.06				0			0	0			0	0	0	1.12	0	0	0
20120322WS-5a	19	2.4	1.71	0.52	**3.33**	2.7	2.25	4.7	0.37	0.55			2.1	10	0.34			7	14.9	3.2	10	
20120322WS-5b	3	0.4	0.25	0.14	**7.41**	5.5	0.49	0.7	0.107	0.08			5.5	10	0.04			3	1.9	4	3	
20180718WS-C	1,332				19.52				0			0	0			0.16	796	9.12	1.76	0	0	1,417
20150622LC-1C	788	0.52	0.28	0.36	**33.85**	1.3	1.42	<0.2	0.309	0.09	0.014		2.6	<10	0.05			5	2.7	3.7	200	
20180718WS-I	0				0.15				0			0	0			0	0	0	0	0	0	0
20180718WS-E	125.98				27.74				0			22,982	0			0.80	1,082	2.16	2.28	0	0	1,659
20180718WS-D	359.8				58.83				0			0	0			1.28	2,054	12.58	36.28	0	0	360
20140926WS-A	30	0.78	0.18	0.36	**62.32**	3.2	1.87	0	0.035	0.1			0.7	0	0.03			15	2.1	2.3	79	
20120322WS-6	2	3.64	2.18	1.08	**13.53**	20.1	3.93	2.6	0.016	0.77			49.1	10	0.33			1	8.6	33.5	150	
20140624WS-C	9	2.6	1.42	0.89	**5.92**	15	3.33	2.2	0.01	0.53			15.2	10	0.22			0	7	18	176	
20140926WS-B	115	2.41	1.31	0.7	**27.63**	10.7	2.82	2.5	0.019	0.49			10.2	20	0.2			8	6.5	11.9	200	
20150923LC-2	165	0.95	0.75	0.12	**12.83**	2.7	0.69	0.9	0.019	0.24	0.052		5	<10	0.14			21	3	3.3	3	
20150923LC-3	280	3.32	2.22	1.04	**5.19**	16.3	3.41	4.3	0.041	0.71	0.028		37.3	<10	0.4			54	15.4	27.1	30	

2019-7-25 Sunrise Chemistry	**p. 4**																					
2019 XRF & previous outside analyses																						
	Pb	Pr	Rb	Re	S	Sb	Sc	Se	Si	Sm	Sn	Sr	Ta	Tb	Te	Th	Ti	Tl	Tm	U	V	W
Sample ID	ppm	ppm	ppm	ppm	ppm	ppm	ppm	ppm	%	ppm	ppm	ppm	ppm	ppm	ppm	ppm	ppm	ppm	ppm	ppm	ppm	ppm
Sunrise-O1	0	0	0		2,230	11.64		0	8.49		36.32	60	0			0	0			0	920	0
Sunrise-O2	0	0	0		2,261	28.16		0	6.94		29.1	20	0			0	0			0	948	0
20190725GF-1	0	0	2.8		2,651	9.4	296.6	0	7.57		29.4	128	0		193.1	0	741.4			0	1,169	0
20180718WS-J	0	0	0		854	0		0	2.64		32.68	320	0			0	0			0	684	0
20180718WS-F	0	50.6	0		263	117.2		4.6	0.48		424	126	258			0	6,697			0	2,047	0
20140624WS-B	5.9	0	82.6		0	37.4		19.8	5.60		231	118	172			4.8	14,899			3.0	2,795	117
20140926WS-C	0	2.67	53.3			3.68	13	1.3		3.58	1	359	0.3	0.53	0.01	4.57		0.09	0.23	4.74	191	27
20150622LC-1B	152	2.65	2.2	<0.001		6.57	1	0.9		5.42	<1	708	<0.1	0.72	0.1	0.51		0.13	0.33	11.05	397	33
20180718WS-A	278	20.5	8.7		951	100		38.7	6.54		266	3,909	235			0	10,763			15.3	2,562	52.3
20120322WS-4	5	2.1	41.9			1.09	9	0.8		2.93	0	377	0.1	0.61	0.04	2.1		0	0.26	4.1	251	3
20120322WS-3	4	2.83	111.5			0.79	10	0.2		4	0	969	0.3	0.58	0.02	3.56		0.7	0.24	3.18	180	3
CC136B	38	1.66	2.1			0.56	2	0.2		2.66	<1	20.5	<0.1	0.37	<0.01	0.25		0.07	0.06	2.56	341	4
CC135	25	0.46	0.8			0.39	1	<0.2		0.81	<1	20.7	<0.1	0.13	0.01	0.13		<0.02	0.06	3.33	754	3
CC134	50	64.9	45.8			0.34	3	0.4		33.1	1	6,450	0.4	1.3	<0.01	5.54		0.23	0.23	3.9	166	2
20120322WS-1	0	2.1	223			0.12	6	0.2		5.15	0	1,875	0.1	0.48	0.19	1.84		1.2	0.09	0.75	157	0
20150622LC-1A	0	0.11	0.6	0		0.57	0	0.2		0.15	0	32.4	0	0.01	0	0		0	0	3.9	11	1
20150622LC-1Aa	8.0	124	2.4		0	16.8		10.8	28.29		67.6	1.8	134			2.4	26,488			7.3	3,056	27.0
20150622LC-1Ab	15.3	0	4.0		203	21.8		2.9	24.43		73	38.4	172			2.9	23,251			6.5	2,422	22.4
20180718WS-B2	16.1	0	2.3		241	24.9		1.0	24.41		72	1,401	56.8			2.2	16,128			15.2	2,025	42.0
20180718WS-H	0	0	0		0	5.84		0	44.50		0	0	0			0	0			0	0	0
20180718WS-G	3.17	0	4.15		0	5.1		0	9.27		9.23	30	0			0	0			4	0	0
20180718WS-B	0	0	0		255	0		0	42.50		0	254	0			0	0			3	0	0
20120322WS-5a	23	0.69	6.5			6.83	3	0		1.56	6	24.2	1.2	0.38	0.42	7.45		0	0.28	16.35	77	8
20120322WS-5b	349	1.07	7.4			8.7	1	0		0.7	0	129.5	0.1	0.06	0.25	1.55		0	0.03	5.45	90	2
20180718WS-C	20.12	0	0		877	12.3		0	31.88		15.7	475	0			0	0			0	0	0
20150622LC-1C	36	0.6	0.7	0.001		4.56	1	0.3		1.39	<1	258	<0.1	0.13	0.01	0.06		0.86	0.04	9.36	174	11
20180718WS-I	0	0	1.26		95.46	0		0	43.98		0	10.92	0			0	0			20.98	0	0
20180718WS-E	18.28	0	109		1,358	8.96		0	13.95		11.86	618	0			0	0			0	507	0
20180718WS-D	0	0	0		672	0		0	7.73		31.98	0	0			0	0			0	873	0
20140926WS-A	0	0.3	0			5.09	0	0.4		1.2	0	615	0	0.24	0	0.57		0	0.03	2.97	37	41
20120322WS-6	27	9.63	148			1.36	11	0		5.64	2	197	0.8	0.59	0.08	19.85		0.8	0.33	3.89	104	3
20140624WS-C	0	4.29	126			0.52	20	0.6		3.9	2	118.5	0.5	0.41	0.01	6.52		0.06	0.24	1.48	82	1
20140926WS-B	26	3.01	61.4			3.25	7	5.5		2.8	1	88.9	0.8	0.41	0.07	7.48		0.05	0.18	39.3	92	7
20150923LC-2	16	0.92	7	0.016		9.24	1	5.5		0.55	6	8	0.1	0.14	0.07	1.37		0.39	0.12	11.2	125	5
20150923LC-3	83	7.8	184.5	0.003		1.07	10	6.9		4.73	2	164	1.3	0.55	0.05	13.65		0.76	0.39	9.71	421	4

2019-7-25 Sunrise Chemistry				p. 5	
2019 XRF & previous outside analyses					
	Y	Yb	Zn	Zr	LE
Sample ID	ppm	ppm	ppm	ppm	%
Sunrise-O1	0		3,820	0	23.17
Sunrise-O2	0		1,216	0	20.38
20190725GF-1	0		3,033	8.3	30.26
20180718WS-J	0		416.24	0	46.23
20180718WS-F	0		83.7	5.0	39.22
20140624WS-B	12.3		632	93	67.12
20140926WS-C	19.4	1.46	1160	33	
20150622LC-1B	26.2	1.85	264	16	
20180718WS-A	61.4		502	32.4	61.14
20120322WS-4	20.7	1.69	715	26	
20120322WS-3	17.2	1.61	867	41	
CC136B	5.5	0.4	10	5	
CC135	3.8	0.34	3	11	
CC134	18.2	1.49	13	129	
20120322WS-1	5.8	0.67	1,830	36	
20150622LC-1A	0	0	10	4	
20150622LC-1Aa	1.1		77	3.6	67.74
20150622LC-1Ab	1.9		86	3.5	71.80
20180718WS-B2	5.4		59	10.7	72.16
20180718WS-H	0		0	0	55.45
20180718WS-G	3.53		86.4	2.38	59.97
20180718WS-B	0.52		0	0	57.35
20120322WS-5a	16	1.94	10	172	
20120322WS-5b	2.6	0.23	4	24	
20180718WS-C	0		138.72	0	47.72
20150622LC-1C	2.1	0.29	636	4	
20180718WS-I	5.4		0	70	55.83
20180718WS-E	6.6		32.52	74	50.22
20180718WS-D	0		0	0	30.54
20140926WS-A	2.4	0.15	23	2	
20120322WS-6	24.2	2.05	185	88	
20140624WS-C	15.2	1.58	286	81	
20140926WS-B	15.6	1.07	248	104	
20150923LC-2	6.5	0.89	15	30	
20150923LC-3	20.1	2.68	162	132	

Chapter 7

Powars II Projectile Points and Bifaces

Bruce Bradley
and Greg Nunn

Introduction

This chapter focuses specifically on the sample of flaked stone artifacts identified as either bifaces (preforms) or projectile points recovered from the vicinity of the Powars II site over the past decade. Additionally, it is restricted to the points that have previously been identified as Clovis, Goshen, Folsom, Unfluted Folsom, Plainview, and Midland. Later point types including Agate Basin, Hell Gap, Alberta and Frontier Complexes are not included. This decision was based on the concerns surrounding the possible identification of a new point type, Sunrise, which is clearly not an issue with these later types. We have also chosen not to engage in lengthy discussions about what this new type might mean in relation to the Goshen-Plainview-Folsom-Midland debate. There are numerous detailed published discussions spanning the past quarter century (e.g. Frison et al. 1996:205-215; Holliday et al. 2017:285-292). Our endeavor has been primarily descriptive with emphasis on the nuances of detailed technological and geometric morphometric analyses. We also don't engage in a retrospective discussion of the issues surrounding the establishment of a new type, nor the issues of splitting or lumping. Our basic philosophy is, until one identifies and establishes the underlying evidence, it is better to split. Then based on theoretical approaches, categories may be lumped. We also understand that artifact types are modern constructs and do not necessarily represent concepts of the people who made and used them.

Sunrise Point Type

This description is based on the characteristics of the bifaces and purported Sunrise points included in our sample from the Powars II site (48PL330). Classification as Sunrise points was based initially on a set of technological attributes that cluster and are different, as a group, from those exhibited by other point types; specifically Clovis, Goshen, Plainview, Folsom, Unfluted Folsom and Midland. Secondarily, the range of sizes and forms of the point types, from Powars II and elsewhere in the region (see Appendix 7-1), were examined through clustering by geometric morphometric analyses (Hauser this volume). Combined, technology and morphology account for our proposed type description. In accord with the traditional way of naming point types, the type is named after the town of Sunrise, Wyoming in which the Powars II site is located. Technology proved to be the deciding factor as sizes and forms of Sunrise points had significant overlap with other point types. There are many terms used to describe morphological and technological features of flaked stone. We follow the definitions published by Bradley and Frison (1996:43-45) and Bradley et al. (2010:56-106).

Our analyses are presented in Appendices 7-3-1 through 3-4.

Sunrise Production Technology

Raw Materials

A range of stone types was used to make the bifaces and Sunrise points included in our sample. The bifaces are of local origin (area of the Hartville Uplift), including the recently identified source named JV chert, which occurs close to the ochre mine. However, several of the Sunrise Points are of non-local stone, specifically Knife River 'flint' and Powder River porcelinite. These sources are substantially northeast and north, respectively, of Sunrise. In terms of flaking quality, a wide range is represented, from medium to very fine. Selection of high-quality stone does not seem to have been a priority. Nevertheless, the stone needed to be fine enough to allow the use of pressure thinning. We identified no clear evidence of intentional heat treatment.

Blanks and Preforms

The vast majority of Sunrise points exhibit no evidence of the initial blank type, with only the pressure thinning scars remaining, with a couple of exceptions. However, there are two likely flake blanks (Figure 7-1.) and many percussion shaped bifaces in the sample along with other early point types, including all of those mentioned above, plus Agate Basin, Hell Gap, Alberta, Cody and some unidentified. Based on the sizes, morphology and technological attributes we have attributed some of the biface blanks to specific point types, but many, including these flake blanks, could have been made into any of the points of interest, with appropriate flaking techniques. It is also evident that the flake blanks and few of the bifaces were not large enough to be made into 'average' Agate Basin, Hell Gap, Alberta, or Cody points, with two notable exceptions (Figure 7-2.). This indicates that if the bifaces were point preforms, they would be from pre-Agate Basin cultural uses of the site.

There are different ways to describe the process of the production of a biface, mainly designed around perceived changes in intended outcomes through the process. These have historically followed the concept of staging starting with the raw material and proceeding through a set of technological stages (Callahan 1979). However, we use terminology designed to identify specific goals of biface production as proposed by (Bradley et al. 2010:77-78) rather than generalized technological stages. This allows descriptions to be specific to specific assemblages.

We use a three-phase system; early, middle and late. On occasion, these are subdivided into subphases such as early/middle, late/early, etc. Keep in mind that these are arbitrary divisions of a continuous process and not technological or morphological predetermined stages. There are no exact descriptive criteria that determine when one phase ends and the next begins (an interval). Different reductions may have significantly different proportions of effort in the phases depending on many factors specific to the individual productions. The knapper decides for each piece when a phase goal has been reached, and as knappers we have made these decisions for our biface sample. We have also tentatively assigned many of them to specific point types. Images of all of the bifaces, their tentative assignments and the data spreadsheet may be found in Appendix 7-2.

Blank Selection/Production

Most finished bifacial pieces do not retain evidence of their initial form, however, occasionally it is possible to determine blank type based on cortex (or subcortex) and/or remnant flake scars of a flake blank.

Unmodified piece (Figure 7-3.) Many stone sources contain small pieces that serve well as biface blanks in their natural form. When evidence is present on an archaeological specimen it is usually exhibited by remnants of natural exterior surfaces such as cortex or subcortex on both faces. If cortex is present on just one face, the original blank form could have been a cortical flake.

Flake blank (Figure 7-4.) When evidence of a flake blank form is present on a biface, it is usually expressed by remnants of the original flake surfaces. When present, these scars may also indicate the orientation of the biface in relation to the flake blank. This is especially the case when the remnant is of the bulbar (ventral) face of the flake blank where the

Figure 7-1. Flake Blanks

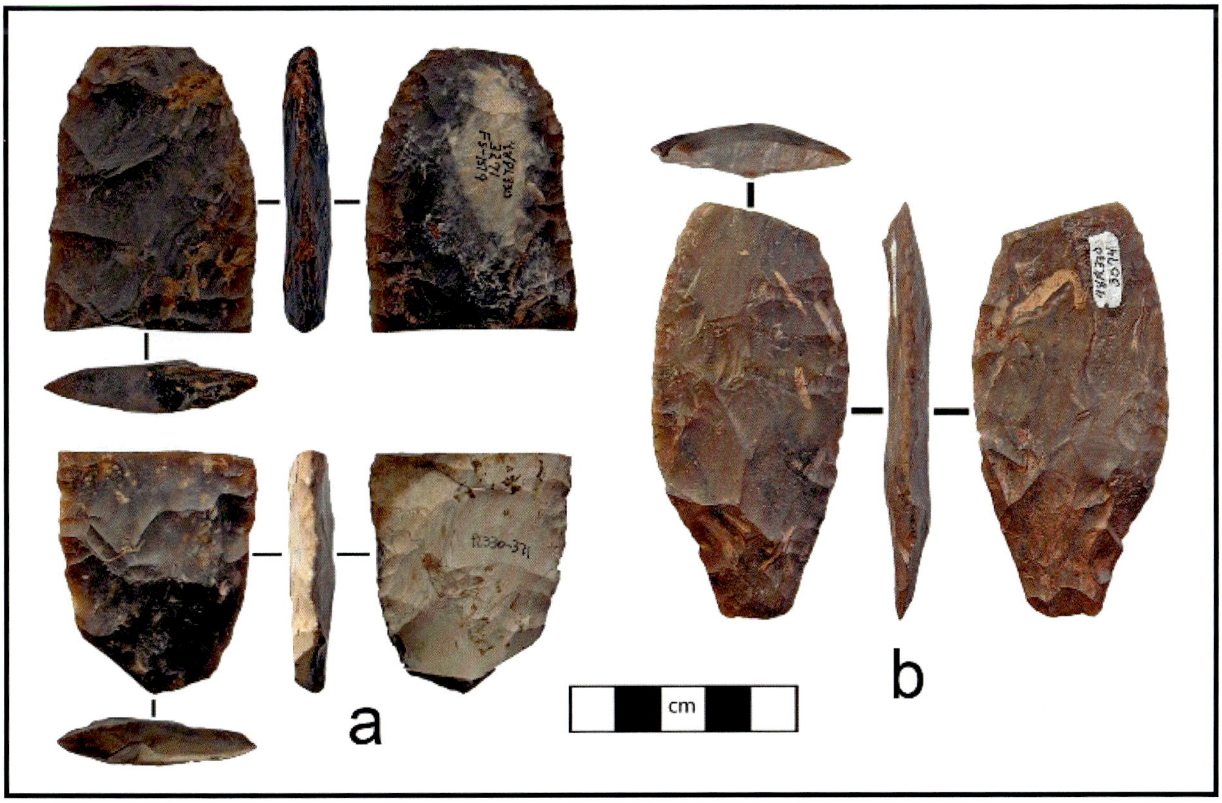

Figure 7-2. Possible Alberta Point Preform

Figure 7-3. Biface Made on Natural Form

Figure 7-4. Early Phase Biface Made From Flake Blank

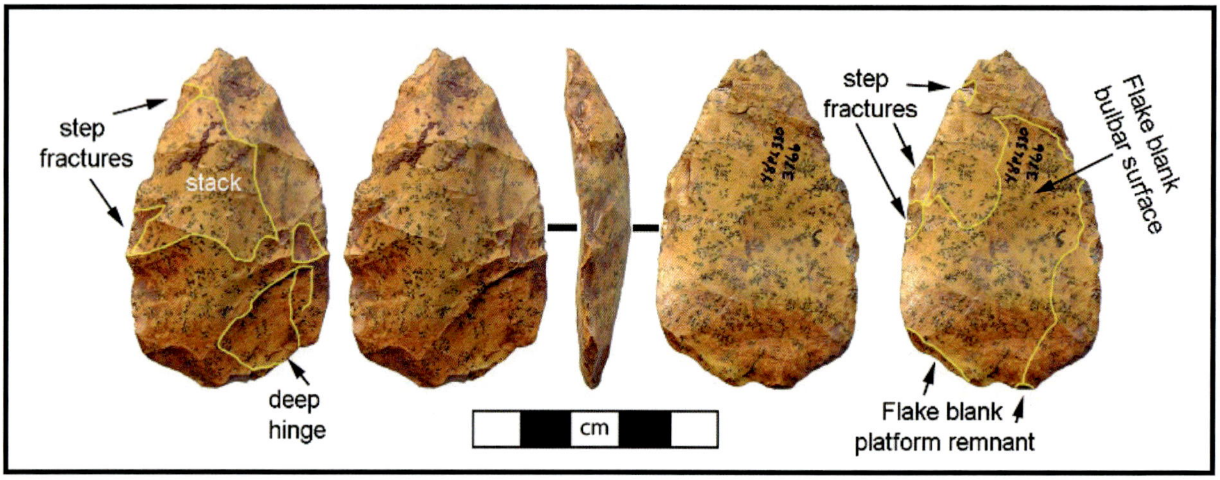

Figure 7-5. Biface Made on End-Struck Flake Blank

flake axis is discernible. While the orientation may be simply the result of the efficiency of producing the biface from a particular flake, it may also be a preference of a particular technological tradition. If there is a strong pattern of a particular orientation (for example selection of side-struck flake blanks) it could indicate a preference for flake core types and thinning strategies. There is also a choice as to how the biface is oriented to a flake blank with the biface base being either at the platform, side or distal end of the flake. This is especially apparent when there is a remnant of the flake blank platform. Another trait to consider is the curvature of the blank. Experience indicates that the best (straightest) flake blanks have hinge terminations.

It is also sometimes possible to infer end-struck flake blanks from the section and cross-section of a totally bifacial piece. If a biface has a plano-convex cross-section and a slightly curved section at the distal end toward the plano (ventral) face, this may indicate the original form was an end-struck flake, with the base of the biface being at the platform end (Figure 7-5.).

One might expect features on bifaces that would indicate the nature of the raw material form, but this depends a lot on the natural forms that were available and selected. It is necessary to understand what forms would have been available for each type of raw material.

Morphological Issues and Challenges of Biface Production

Flaking quality and available forms of raw material highly influence the choices a knapper has and there are several issues that influence reduction choices. Since Paleoindian points tend to be relatively thin, regular, and uniform the following issues need to be addressed during raw material and blank selection or production: size, fracture quality, cortex, inclusions/flaws, thickness, curvature, 'square' and round edges, irregular surfaces, platform remnants and bulbs of applied force. There are also the non-technical qualities that may influence the selection such as color, translucency, visual texture and perhaps even the location of the source on the landscape, which may have been perceived as having special significance.

The Powars II location is at a source of highly variable flakable stone that has been identified as JV (John Voight) chert, agate, chalcedony, jasper and opalite, depending on its opaqueness or translucence and fineness of texture. These are subjective assignments, and all varieties may be present in a single piece of the raw material. In addition, the site is close to the well-known Spanish Diggings and Hartville sources of silicified sandstone (sometimes called orthoquartzite) and various forms of chert and jasper. In this study, we have called opaque, medium quality stone chert, which is a catch-all category, used when none of the other types apply. Agate is translucent with a range of colors. Opalite is agate but extremely fine-grained and glossy. Jasper is opaque with an extremely fine waxy texture. Silicified sandstone is just that. It is sedimentary in origin (quartzite is metamorphic) and comes in many textures and flaking qualities from coarse to extremely fine.

A related issue to raw material is the knapping tool kit that was being used. Unfortunately, this aspect is very poorly understood for PaleoAmerican assemblages because of the paucity of in-place production areas and poor preservation of organic materials.

Biface Production Options

Many flaking methods (Bradley et al. 2010:64-79) are apparent in the Powars II bifaces including:

Non-thinning is where flake removals were abrupt or invasive but did not remove relatively more thickness than reduction in width (Figure 7-6.). This method was often used in the early phase where the primary objective was the establishment of a regular biface plane. It was also applied selectively throughout the process, especially for shaping the outline and regularizing the margins and surfaces, for example removing deltas between primary thinning flake scars. It was also used for platform preparation. Generally, this was done with controlled percussion, but occasionally, especially near the tip and along the stem margins, and for isolating platforms, pressure was employed. This method was used in all PaleoAmerican technologies and is not considered diagnostic.

Standard thinning is where flakes were struck or pushed, which usually ended in feather terminations and removed relatively more of the

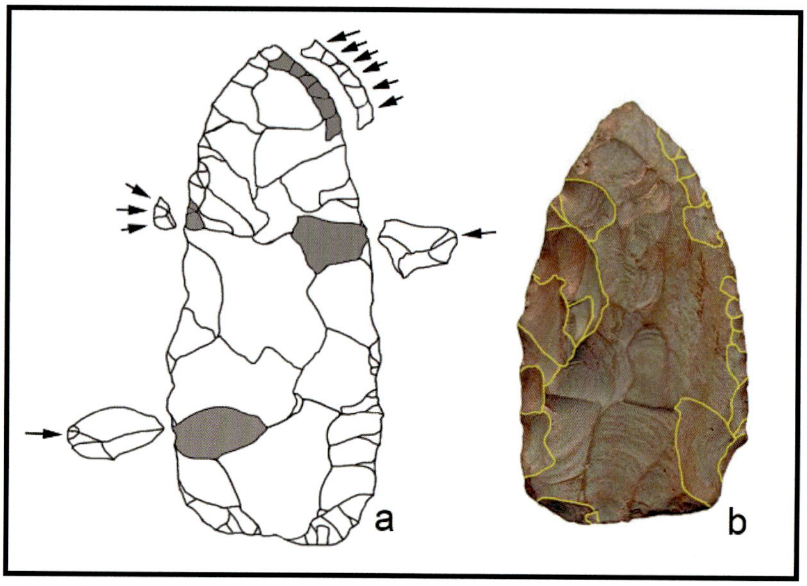

Figure 7-6. Biface Non-Thinning Flaking, a)schematic, b)48PL330-1009

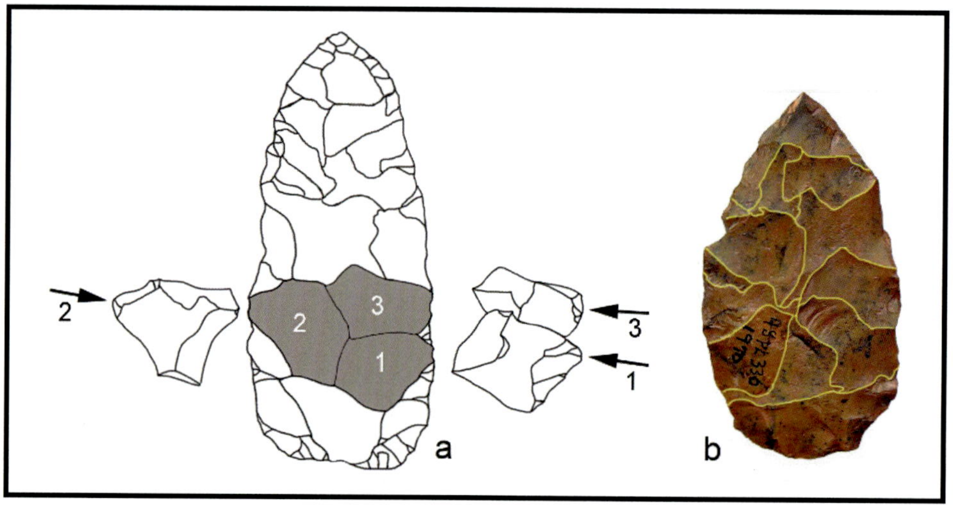

Figure 7-7. Standard Thinning, a)schematic, b)48PL330-1970

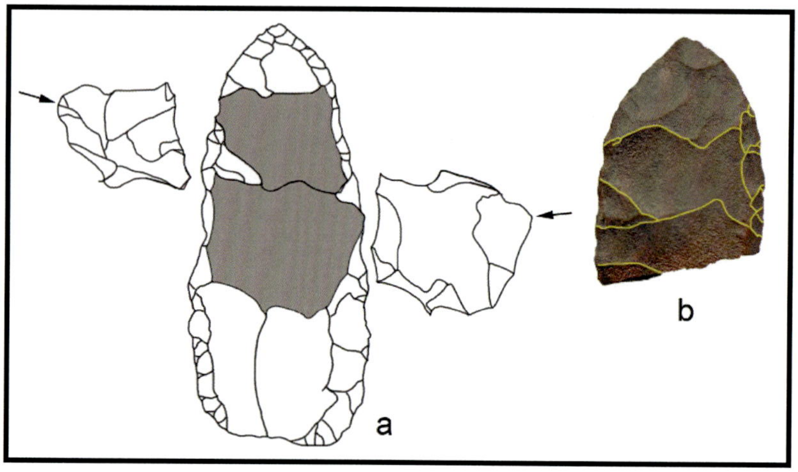

Figure 7-8. Full Face Thinning, a)schematic, b)48PL330-1169

thickness than the width (Figure 7-7.). When applied from two opposing margins this usually produces an overlapping scar pattern (Bradley et al. 2010:Figure 3.10). Most Powars II bifaces were produced with percussion, which is not considered diagnostic.

Full-face flakes travel most of the way across the biface, terminating at or near the opposite margin (Figure 7-8.). Full-face flakes are an efficient method of thinning a biface and when widely spaced produce flat cross-sections. While present, this method does not occur with enough frequency to be considered a primary approach used by Powars II knappers and is therefore undiagnostic.

Overshot flaking extends across the entire face and removes a portion of the opposite margin (Figure 7-9.). This may happen unintentionally and remove too much of the margin (plunging termination) resulting in abandonment. However, when controlled, this technique is a highly effective way to thin and flatten a biface (Bradley et al. 2010:68-77). To infer that controlled overshot flaking was being used systematically in a technology, there must be a significant proportion of successful overshot flakes and/or flake scars. Some bifaces exhibit overshot flake scars and when present with other features including remnant square edges, full-face flake scars, complex platform preparation and/or end-thinning, we have identified them as Clovis.

Diving flaking allows optimum biface thinning when applied to both faces. This method is initiated when a thinning flake is set up and removed so that it intentionally terminates in a hinge or step fracture near the midline of the biface (Figure 7-10.). This is usually accomplished by preparing a highly projected, strong platform (Bradley et al. 2010:66-67, Nami 2014:198). The flake is struck with an inward angle and a sharp blow, without follow-through, so that it either hinges or steps. If overstruck, it may produce an overshot removal or a plunging termination resulting in abandonment.

Once one diving flake is removed it is easier to produce adjacent flakes so that they also end in hinges or steps. These terminations are then removed by striking equivalent flakes from the opposite margin, resulting in a depressed area near the biface midline. When carefully and systematically applied to both faces, this method is extremely effective at thinning.

This method of thinning is difficult to control and depends on careful platform preparation and adjustment to the biface plane. Archaeological evidence of platform isolation has been noted with isolated platforms indicating that many were carefully prepared individually, including grinding (for a discussion of complex platforms see Bradley et al. 2010:65-67). Diving flaking is best known for Folsom ultrathin knives but may have been employed in the middle phase of preform production for Goshen and Folsom points. This is present in the Powars II sample and used to suggest a Goshen affiliation.

End thinning is where normal or diving flakes were removed from the end of a biface during the middle phase as a method of thinning the base, or on occasion the tip (Figure 7-11.). These do not necessarily result in overall thinning (based on w/th taken at maximum width). End thinning flakes usually have flake scars at approximately a 90o angle to their axis, as with channel flakes, but the scars tend to be larger and more irregular. The flakes are also more asymmetrical than many channel flakes. In many cases it is difficult to distinguish them from standard thinning flakes. Basically, they are the same, just removed from one of the ends. End-thinning is present on many of the Powars II bifaces and alone is not considered diagnostic.

Fluting is basal end thinning accomplished in the late or final phase where a significant portion of the scar remains visible on the finished piece (Figure 7-12.). When fluting was considered successful only one scar was produced, however, on occasion the first attempt was considered inadequate (too narrow, off-center, too short, etc.) a second and sometimes a third flute or thinning flake was removed. These were either parallel and adjacent with or directly beneath the first flute. Channel scars are absent on the Powars II bifaces and rare on finished points, an exception being the Colby site Sunrise point. This may primarily be the result of post-use reworking that eliminated the channel scars on the finished points. However, based on the sizes of the recovered bifaces, one can

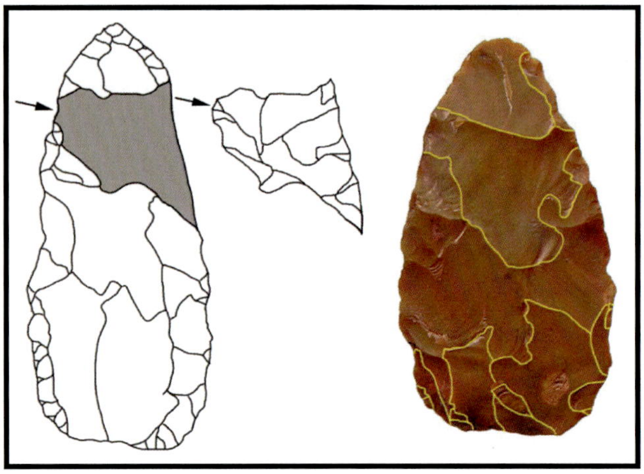

Figure 7-9. Overshot Flaking, a)schematic. b)48PL330-1120

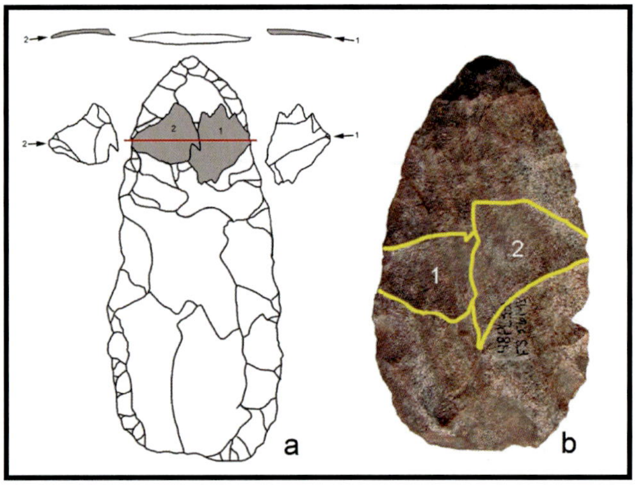

Figure 7-10. Diving Flaking, a)schematic, b)48PL330-Fs561

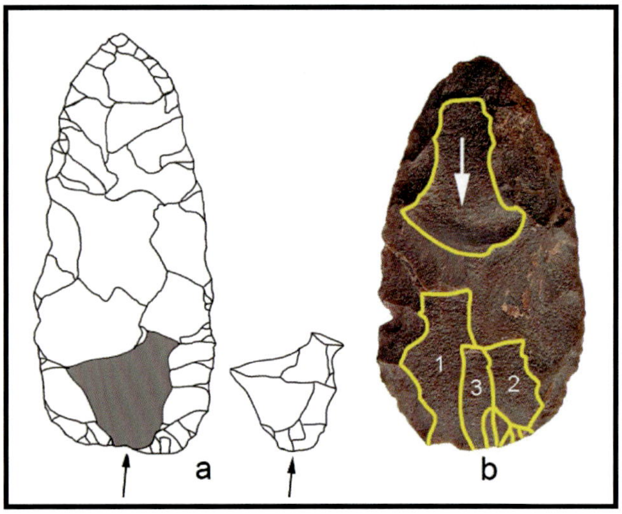

Figure 7-11. End-Thinning, a)schematic, b)48PL330-561-B

make a case that many of the recovered points are in their original, non-reworked state, indicating that fluting was not a standard part of the production technology of Sunrise points.

Flaking Patterns

Bifacial flaking produces many different visual impressions based on flake scars. These may form distinctive patterns or be nonpatterned. These are controlled by the knapper by flake sequencing, spacing and type of force application. Generally, the Powars II bifaces do not exhibit particular flake scar patterns, with most having what we call overlapping. This is where flakes are removed from opposite margins in a selective sequence (not random) and the scars interfinger at variable locations on the face of the piece. We also recognize comedial, diagonal, diving, end-thinning, full-face, marginal abrupt, marginal invasive, overshot and unmodified. Whichever one of these dominates a face is how it was classified. Keep in mind that flake scar patterns may develop during the process. For example, a middle phase biface may exhibit overlapping flaking but go on to receive a final diagonal serial sequence. There are a few flake scar patterns that we consider diagnostic, such as overshot (Clovis), percussion comedial serial (Alberta), pressure comedial serial (Goshen), and overlapping diagonal pressure (Frontier Complex). Sometimes these patterns are most evident in the early and middle phases (end-thinning and overshot) and are eliminated by further flaking, while others (comedial and diagonal serial) only begin to develop in the late/middle to late phases.

Condition

We use three categories for condition. A fragment is a piece where it is difficult to estimate what the missing portion size and or shape would have been. Incomplete is where a small piece is missing but its size and shape can be reasonably estimated. Complete is where nothing is missing.

Measurements

Each biface had dimensions recorded in millimeters to one decimal. Length was the maximum dimension along the section. For fragments, length was taken at the maximum dimension. Width was taken at the maximum dimension perpendicular to length. Thickness was taken at the maximum dimension at the width. Weight was recorded in grams to one decimal (Figure 7-13.).

Abandonment

To try to understand why the bifaces were abandoned, we assessed any obvious reasons why they could not have been finished into a point within the Powars II assemblage. Some were obvious, such as breaks. Even so, a resulting fragment could still be large enough to be made into a point, but for some reason they were not finished. Whether or not some flaking errors were the reason for abandonment was less obvious. These most often consisted of deep step or hinge terminations that would have been difficult to fix with the current biface morphology (including relative thickness) and flaking expertise exhibited. What a novice would discard as unviable, an expert may have continued. So, we used our judgment as expert knappers to determine whether or not each biface was still viable for finishing. We also considered the dimensions, especially thickness in relation to the remaining width and the finishing technology exhibited on the completed points.

Most of the Sunrise points were finished with thinning pressure flaking. This process reduces the width while reducing the thickness, resulting in higher width to thickness ratios (relatively thinner pieces). This is what we came up with:

Of the late phase bifaces only 48PL330 FS-645 was unlikely to have been completed into a point as it is too relatively thick. It is also a fairly coarse silicified sandstone that would not allow pressure thinning. However, it is still large enough that with expert percussion thinning, there was still some potential.

Six have potential to become Sunrise points (Figure 7-14.). Even the smallest could have been made into points the size of some recovered (Figure 7-14. d-e).

While most bifaces were abandoned because of obvious faults, inclusions or flaking errors, a few may have been completed cutting tools, but some seem to have simply been unfinished. It may be another attribute, specifically the width to thickness ratio, that caused the abandonment. All together finished Sunrise points had w/th ratios of 4.1 to 6.4

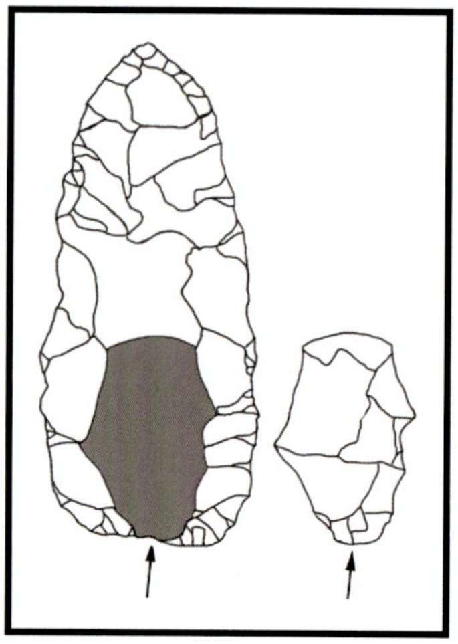

Figure 7-12. Late Phase Fluting

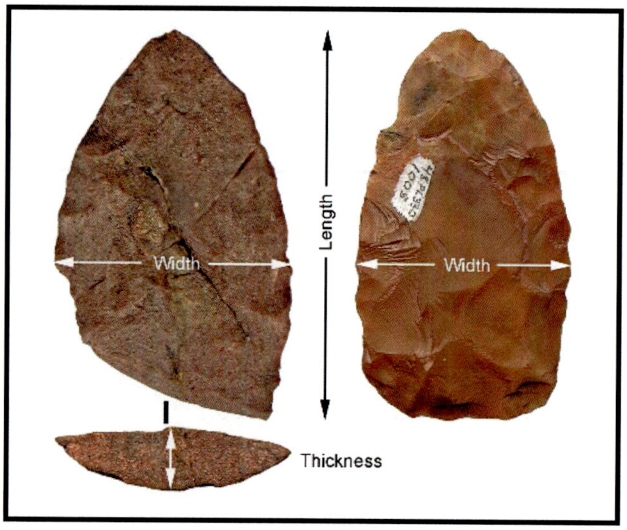

Figure 7-13. Dimensions

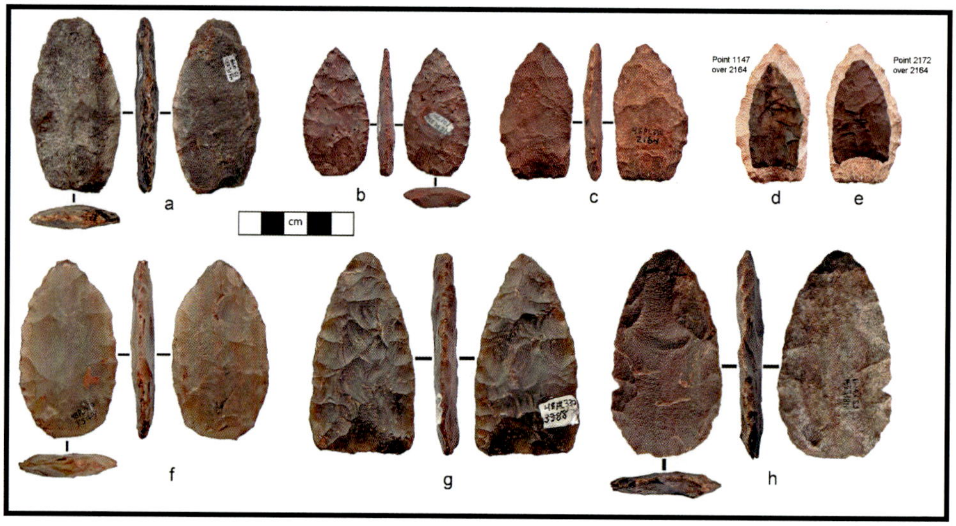

Figure 7-14. Bifaces with potential to be made into Sunrise points, a FS245, b 1017, c 2164, d 2164 superimposed by point 1147, e 2164 superimposed by point 2172, f 1207, g 3388, h 561B)

with a mean of 5.0, while bifaces with unknown abandonment had ratios of 2.5 to 6.0. All but three of them could have been used for any of the point types of interest. However, many of the bifaces were considerably longer than would have been expected for them to have been destined to become Sunrise points. This certainly begs the question as to what they represent.

Point Sample

All of the points from Powars II that were used in this study are illustrated in Appendix 7-1. This includes the analysis spreadsheet as well as individual point observations. The following technological discussion is focused on the 34 points we have classified as Sunrise, based on technological details and general form.

Sunrise Point Manufacture Technology

Shaping and Thinning of Preforms

The standard process used to produce Sunrise points was to make a fully bifacial preform using standard thinning and shaping by percussion. Then they continued to be formed by what we are calling pressure thinning/flattening. This is where pressure flakes are removed with significant inward force so that they travel past the midpoint or thickest area and flatten the face (Figure 7-15a.). These flakes frequently terminated in shallow steps. This type of flaking is also present on Folsom preforms and Goshen points but is generally lacking on Clovis and Plainview points. Pressure shaping was also selectively used with flakes travelling in a curve across the face and not reaching the midline or thickest portion, creating a convex face (Figure 7-15b.). Terminations tended to feather. This was the method typically used to shape Plainview points.

On Sunrise points the pressure thinning flake removals were selective without a specific pattern, and relatively widely spaced. There was a tendency toward removing adjacent but not overlapping flakes and then the space in between them (Figure 7-16.). Folsom and Goshen points tend to have sequential serial flake removals. Platforms for the pressure removals were prepared with small abrupt beveling so that the resulting scars produced thin, sharp edges leaving a 'hollow-ground' effect. There was also a tendency for wide platforms to be removed, possibly indicating the use of a soft pressure tool such as bison horn. Spacing of removals often left distinctive triangular areas (deltas) between them (Figure 7-17.). Sections of a margin would be prepared rather than individual platforms. Light abrasion may have been added to strengthen the platforms. The result was wide, straight, shallow pressure flake scars, normally at a right angle to the midline. We could discern no pattern in the sequence of margins or faces except for the basal thinning, which tended to be done last, although lateral flaking removing basal scars is also common.

Some basal thinning scars are large enough to be identified as fluting (Figure 7-18.), but this is a rare occurrence with only four bases (of 68 possible) exhibiting this feature. It is of course possible that end thinning had occurred during the sequence, with the scars removed by subsequent flaking. However, fluting was not systematically used.

Finishing Retouch

Perhaps the most distinctive technological feature of Sunrise points was the general lack of finishing retouch, unlike that seen on Clovis, Goshen, Folsom, Midland and Plainview points. In flaked stone descriptions the term retouch is used in several ways and inconsistently. In this analysis the term retouch has a very specific definition following Bradley and Frison (1996:44). Retouch is the final edge adjustment flaking, which was done to straighten, dull or sharpen the margin, but not to reshape the tool surface or to serve as a platform for subsequent flaking. On Sunrise points edges were left 'wavy', commonly exhibiting deltas, and are very thin and sharp. The paucity of retouch also means that sections of platform preparation may remain and could be mistaken for retouch (Figure 7-19.). Retouch definitions and Sunrise points with retouch or remnant platform preparation are included in Appendix 7-3.

Margin Dulling

Lower margin abrasion is present on many points but tends to be very light, with only occasional moderate to heavy dulling. Basal abrasion is generally lacking.

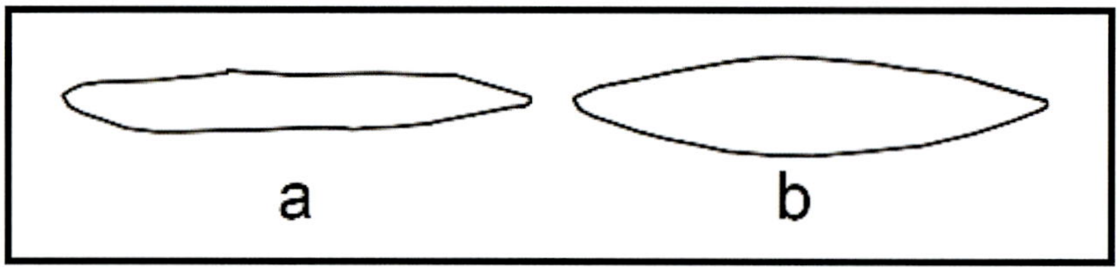

Figure 7-15. Expected cross sections, a pressure thinning/flattening, b pressure shaping.

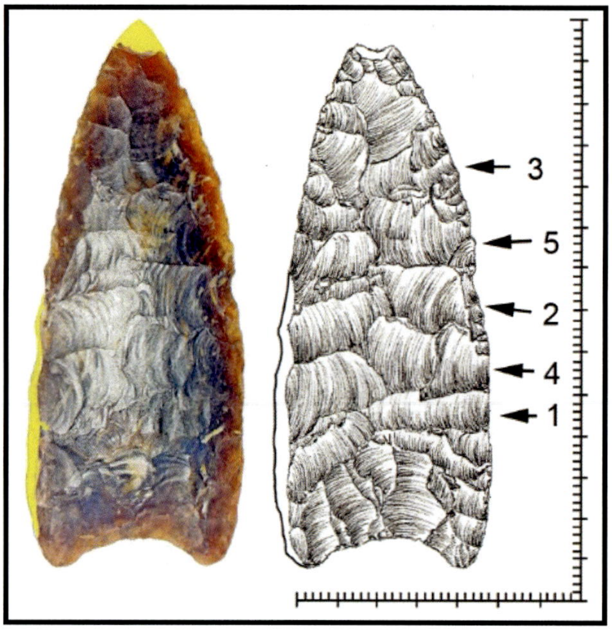

Figure 7-16. Pressure flaking sequence on some Sunrise points.

Figure 7-17. Remnant deltas indicated by yellow shading.

Comparisons

Many of the points from Powars II have previously been identified as Clovis (Frison et al. 2018) or suggested to be Sunrise-Clovis (Spencer et al. 2022). We consider neither to be appropriate for our sample. There is little technological similarity to classic Clovis, even Clovis points that have been reworked (e.g. Haury et al. 1959 Figs. 12e and 12h; Huckell 2007:199-201). The distinctive pressure thinning is lacking on Clovis, and fluting was not a standard application during Sunrise point production. There is also little evidence that Sunrise points were made with pressure directly on flake blanks, such as present on many Plainview points. Occasionally, flat pressure flaking is observed on Plainview points, but this was usually the result of flake removals from flat flake blank surfaces (Knudson 2017:60 Fig. 2.29b) rather than actual thinning or flattening. Goshen finishing included systematic, patterned pressure thinning/flattening and nearly continuous edge retouch on all margins (Bradley and Frison 1996:43-50).

Assuming that the fully bifacially percussion flaked bifaces in our sample represent the usual Sunrise point production process, this may be an indication of a technological relationship to Clovis in the early phases of production. Nevertheless, the closest overall technological similarity is to Goshen.

Point Morphology

Size and shape are critical aspects of point type descriptions, and this is where general subjective descriptions have frequently fallen short. A current method is called geometric morphometrics (GMM), which attempts to standardize the process (see Hauser this volume). While our goal hasn't been to revisit the issue of type distinctions between Clovis, Goshen, Plainview, Folsom, and Midland, we did consider it important to do a comparative study of morphology to see if our classification of Sunrise points using technology would correlate with morphology. The results were mixed. While there is a significant range of forms in each of the types, there are some clear clusters as well as significant overlap. Neil Hauser did all of the GMM analysis (see Chapter 8 for a detailed discussion). He ran multiple groupings using a wide range of metrical and statistical methods. Ultimately the most useful results were obtained by creating 180 parameters for each projectile point that are the distances from the centroid of the point to the edge of the projectile point at each degree of arc starting at the tip of the projectile point, zero degrees, to 180 degrees from the tip, ideally the center of the base. This resulted in 180 lengths describing each projectile point. The projectile points were assumed to be symmetric so only 180 degrees of arc were used.

Principal components analysis (PCA) describing 95% of the variation was then performed on the 180 parameters after applying standard scaling. The 180 parameters were scaled to ensure each landmark along the edge of the point had the same influence in determining the principal components; that is they had values over the same range of magnitude. In the case of the projectile points used in this investigation, this resulted in two principal components. Those principal components were then used to create a dendrogram (Hauser Figure 8-2.) to visualize the level of sameness between the projectile points. Agglomerative Hierarchical Clustering (AHC) was used to determine the membership of the number clusters indicated at the various levels of sameness (horizontal line). It was subjectively decided that five clusters were the best description of the projectile points. This is also evident when the principal components are plotted (Hauser Figure 8-1.2.), PC1 vs PC2, and the five clusters indicated in that plot.

The results of the clustering of the points used in the sample show some interesting results (Figure 7-20.). Clovis points are found in all clusters, but in greatly varying proportions. In Cluster 1 they occur with a couple of Goshen points but none of the other types. This is basically a Clovis cluster. Cluster 2 is dominated by Plainview points but includes all other types except Sunrise. Cluster 3 is also dominated by Clovis points, but has a large representation of Sunrise points, a few Plainview, but no Goshen, Unfluted Folsom or Midland. Cluster 4 is strongly dominated by Sunrise but with a moderate representation of Plainview and lesser of all the other types. Cluster 5 has too few points to be relevant but those present are only Sunrise and Clovis.

Based on the 5-Group clustering there seem to be significant differences between Clusters 3 and 4 of the Sunrise points. The first is that the

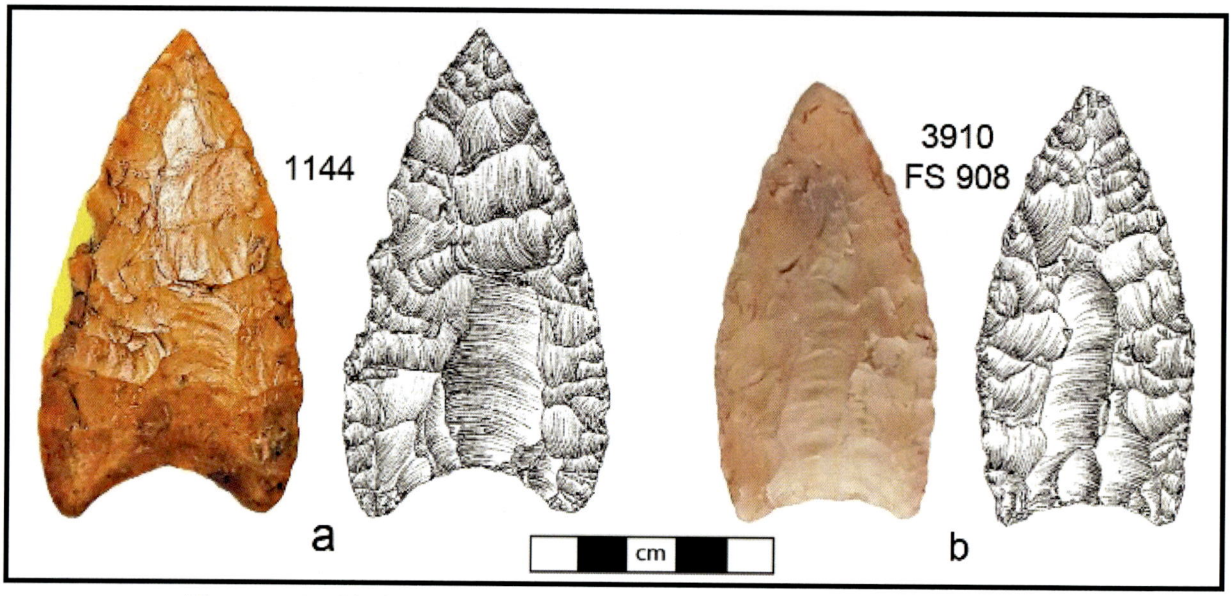

Figure 7-18. Fluting on Sunrise Points, a percussion, b pressure thinning.

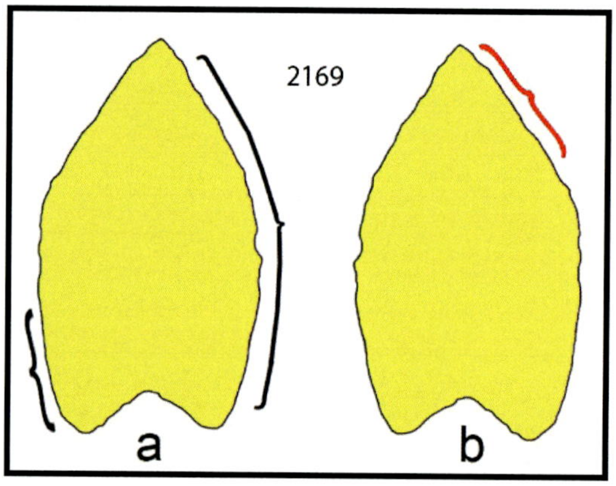

Figure 7-19. Retouch (red bracket), and remnant platform preparation (black brackets) a obverse, b reverse.

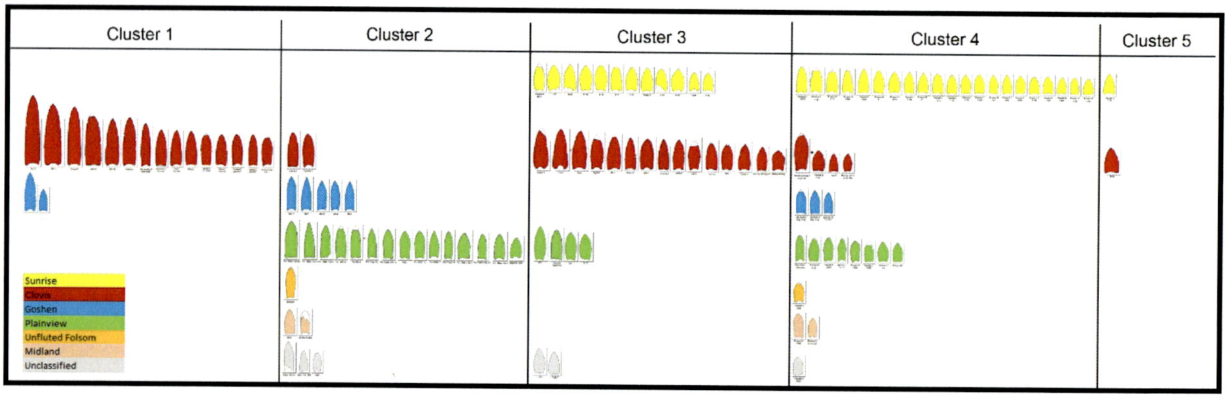

Figure 7-20. GMM 5-group results by cluster. Organized by point type and length, from left to right.

Dimensions (mm)	Cluster 3 (n=11)		Cluster 4 (n=17)	
	Range	Mean	Range	Mean
Max length	43.0-64.0	54.6	31.0-61.0	46.8
Max width	25.5-31.0	28.0	24.0-28.0	25.5
Relative thickness	4.1-6.4	5.5	4.3-5.2	4.7
Length of max width	8.0-30.5	21.0	11.0-23.0	17.0
Base indentation	2.0-7.0	3.8	1.0-5.0	2.8
Weight (grams)	5.7-15.5	9.9	6.2-11.6	7.9

Table 1. Comparison of Sunrise Points Between Clusters 3 and 4.

separation between clusters, including all point types, is evident all the way to the top of the clustering dendrogram (Hauser Figure 8-2.). When metric data is compared between the Sunrise points in Clusters three and four there are discernible differences (Table 3-1). This indicates that the points in Cluster 3 are longer, wider, thinner, have deeper basal indentations, weigh more, and have maximum widths further from the base than points in Cluster 4.

To us, these results indicate that, in terms of shape, Clovis has the widest range of variation, overlapping with all of the other types in this study. Plainview and Goshen forms are also quite diverse. However, Sunrise points cluster strongly, overlapping mainly with Clovis. This is probably why in previous assessments Sunrise points were thought to possibly be reworked Clovis points (Frison et al. 2018) and later suggested to be 'Sunrise Clovis' (Pelton et al. 2022). It has been evident that the points at the Powars II site aren't 'classic' Clovis, and this analysis gives us some insight into the shape distinction of the proposed Sunrise point type, especially when other aspects of morphology and technology are combined. Considering the overlap in shapes of all the types included in our analyses, we recommend that production technology be used as the main distinguishing criteria in defining the included types.

Sunrise Point Distribution

Other than the Powars II sample the only other point that we had available that we consider Sunrise is one of the small points from the Colby site (Frison and Todd 1986, Figure 3.2e). This point (Figure 7-21a.) in GMM Cluster 3 exhibits all the features of Sunrise points except that it also has fluting on both faces and is slightly larger and relatively thinner than average. The other small point from the site (Frison and Todd 1986, Figure 3.2f) exhibits typical Clovis traits; shape, percussion thinning and shaping, fluting, retouch, and lower margin and base abrasion (Figure 7-21b.). Since the Colby Site Sunrise point was reported earlier than the recent publications on the Powars II site, a case could be made to call Sunrise points Colby points but basing the name on this single specimen, which is an outlier of the overall group, seems unhelpful. The two larger points from Colby also exhibit some traits of Sunrise points (Frison and Todd 1986, Figures 3.1a and b), such as some pressure thinning, but lack others. They are atypical Clovis shapes, Clovis sized, show retouch and fluting and have well-abraded lower margins and bases.

Dating

All of the Sunrise and most of the other points recovered from the Powars II site derive from mixed contexts, mainly the result of historic iron mining. Nevertheless, the majority were recovered either on the modern surface or by screening backfill very close to the in-place ochre mine trench. This means that none are in dated associations, but also, they were likely associated with the ancient mining activities starting during or soon after the Clovis era. About all that may be concluded is that Sunrise points were of early Paleoindian origin. If the Colby point is Sunrise, there are associated dates indicating contemporaneity with Clovis (Frison and Todd 1986:22). New dating samples are currently being submitted as part of a project of reanalysis of the Colby materials (Maddie Mackie, personal communication).

Another line of evidence that Sunrise points are not unique to the Powars II site is the presence of non-local stone, specifically Knife River flint and Powder River porcelinite. There are no blanks or preforms of these stones in the collection, so they most likely were made elsewhere and brought to the site. We have also observed Sunrise points in private collections indicating they are present in northern Colorado.

A re-evaluation of the Goshen points from the Midland Complex at the nearby Hell Gap site (Larson et al. 2009) shows that one of them, (Bradley 2009:262-263, Figure 17.4f) may now be reclassified as Sunrise. Otherwise, this type is not represented at Hell Gap.

Conclusions

A main takeaway from this exercise of describing and defining a new point type, is that production technology is of primary importance. Shape (morphology) is also critical, but at least with the Clovis, Goshen, Unfluted Folsom, Midland and Plainview types there are significant overlaps. Technologically, Sunrise points are most similar to Goshen points, but there are also clear differences (Figure 7-22.). Adding a new point type, Sunrise, to the mix may be seen as over-complicating the discussion of what types mean and how they are used to interpret the early assemblages of the High Plains. However, with refining the methods we use to describe and define, as brilliantly demonstrated in Plainview (Holliday et al. 2017), we should be able to significantly expand our ability to interpret this distant, but intriguing, human experience.

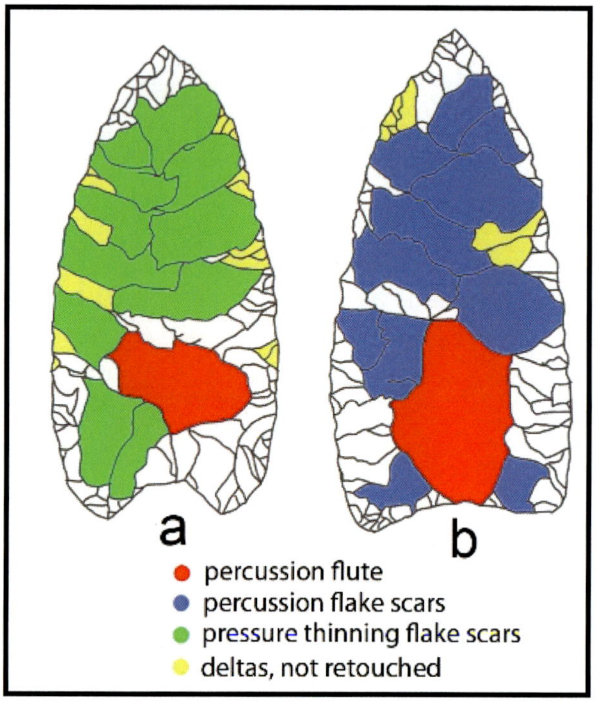

Figure 7-21. Colby site points illustrating different production technologies, a Sunrise, b Clovis.

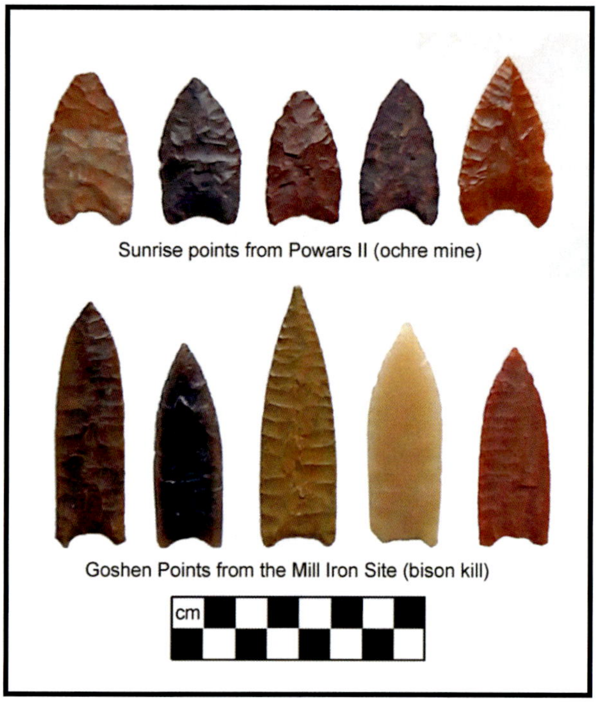

Figure 7-22. Comparison of Sunrise and Goshen points.

Chapter 8

Applying Geometric Morphometrics (GMM)

Neil Hauser

Introduction

Projectile points are typically placed into classes or general shapes composed of several styles. These classes refer to the general shape of the projectile point (herein referred to as points in this chapter). Some common classes used are lance, corner-notched, side-notched, stemmed, bifurcated, auriculate, and triangular. Stemmed are often further broken down into expanding stem, straight stem, and contracting stem. Within these classes, styles are usually based on the geographic area where the projectile points are found, the temporal period from dated sites where they have been found within that geographic area, and the features of the points.

Because there are no hard boundaries between the features in one point style and those in another, the features of the points associated with a style are subject to many factors that can lead to changes in the features, termed "drift," that are associated with a given style. These changes are usually associated with the accumulation of subtle differences in some of the features which lead to changes in the "mental template" a person uses when assigning a point to a style. This may not be unlike what does and did happen, when a modern or prehistoric knapper made a point of a particular style. Over time, it is easy to understand how there might be a continuum of feature values and how one style might evolve into another.

Deciding where to place a boundary on a continuum of feature values lends itself to differences in the assignment of points that have features near those boundaries that define a style. The problem is even more pronounced when considering things like differences in size and features such as whether concave, convex, and flat bases comprise different or the same styles. There is no consensus among archaeologists, which makes using point styles in analysis problematic and can lead to very different conclusions in different analyses and by different analysts.

A set of features might be considered associated with a style, if they occur together in many projectile points. How many points with those associated features is sufficient to call it a style? Certainly, one projectile point isn't enough to define a style, but how many does it take? How many features must be uniquely present to define a different style? How much difference between the values of those features is sufficient to define a different style? These are the kinds of questions that make defining a style difficult and leads to differences in assigning membership to a style.

Deciding which features represent meaningful differences between point styles is subjective and thus subject to variation between analysts, even when archaeologists use accepted typologies, such as Justice's typology for the Southwest US (Justice 2002). Using too many features may result in using random variations that are not representative of the mental template of the knapper. When assigning points to styles is done quantitatively, there are almost always some projectile points included in a style that "don't look like the rest" when assessed by a person. Why? Because each individual assigns different amounts of importance to different features of projectile points when considering which points look similar (Hauser 2021).

To attempt to reduce the subjective decision making in defining styles, geometric morphometrics (GMM) techniques developed by biologists to study shape have been repurposed and applied to points. Typically, in applying GMM to points (e.g., Buchanan and Collard 2010; Buchanan et al. 2019; Smith 2010), several locations, landmarks on the points that define key features, such as base width, length, and neck width, are chosen by the analysts. Then a predefined method for selecting additional landmarks around each point is applied to define the overall shape of each point. The centroid of each point in the dataset is determined using the landmarks, and following the methodology used by biologists. General Procrustes is applied to best fit each point to a template shape. Usually, the template is the average shape of all the projectile points in the study dataset. This reduces the effect of size as a feature and provides a means of comparing just shape. I argue that while removing size is acceptable for some questions, it should be considered carefully before applying it.

Once a set of features or landmarks are determined for each point, principal component analysis (PCA) can be applied to reduce a potentially large number of landmarks or features to a smaller more statistically representative number of independent components. Since each feature or parameter can be considered a dimension, PCA is termed dimension reduction. These components form a smaller, multi-dimensional space in which the projectile points are defined. However, these principal components are a linear combination of the original parameters or features, and therefore are not easily visualized or physically describable. This makes relating the principal components to the actual images of the points less clear.

Finally, cluster analysis is usually accomplished using either the features, parameters, or principal components in one or more mathematical clustering algorithms. These clustering algorithms use various criteria to group (or cluster) those points that are most similar when considering all the inputs (features, parameters, or principal components). A new group is defined when the range of values within an existing group and those of a point exceeds a criterion. With the advancement of data science and machine learning, many new clustering algorithms have been developed and refined. Favorite algorithms for clustering points include Agglomerative Hierarchal Clustering (AHC) and K-Mean clustering.

Powars II Analysis

For this analysis, one hundred nineteen scaled images of Paleoindian points were provided by Dr. Bruce Bradley (see Appendix 8-1.). These included the points of interest plus several examples of the other styles of Paleoindian points that occur in the region, including Clovis, Goshen, unfluted-Folsom, Midland, and Plainview. Each of these points was included in the GMM analysis.

Paleoindian points are, at first look, simple in shape, no notches or tangs. However, upon closer examination, there appear to be features that may or may not have been part of the knapper's mental template for making a particular point style, such as where the maximum width occurred, etc. To capture as many of the features that may have composed the knapper's mental template, the points were described by a set of lengths from the centroid of the point to the edge of the point. Using the centroid is problematic since reworking of the point will shorten the point and result in the centroid moving towards the base. However, with Paleoindian, there is likely no other choice than to use the centroid as the reference location and an effort was made to use points that were not reworked.

To ignore small random and unintentional variations due to flake scars and edge damage, the point shape was approximated by a series of straight-line segments (black lines in Figure 8-1.). When the slope of the edge of the point changed, a new line segment was created by placing a landmark there. The number of line segments describing each point varied and depended on the amount of curvature (or changes of slope) of the edge for each point. The number of landmarks needed to describe the points in this dataset varied from 16 to 37. The open-source software tpsDig2.0 (Rohlf 2004) was used to create a file containing the locations of the landmarks.

To create a high-fidelity description of the point, the length between the centroid of the point and the line segments that approximated the shape of the point was calculated for every degree from the tip (0 degrees) to 180 degrees, nominally the center of the base. Thus, half of the point, from the tip to the center of the base, was defined using 180 lengths. Since symmetry of the point was assumed these 180 measurements became the features used in the GMM analysis. The lengths from the centroid of the point to the edge for each degree were calculated using software written by the author for this analysis. The output of the software, lengths for each degree for each point, was the input for calculating the principal components and clustering analysis.

Once each projectile point was described by the 180 parameters, principal component analysis (PCA) describing 95% of the variation was done. This resulted in two principal components describing each of the projectile points. Those values were scaled to ensure that differences in magnitude between the two principal components would not bias the hierarchical dendrogram and clustering.

In this analysis, hierarchical clustering was used to create a dendrogram of clusters. This approach is straightforward. It starts by finding sets of the two most similar shaped projectile points based on the features, parameters, or principal components and then adds the next most similar group of points until all the points have been included, creating a hierarchical tree (dendrogram). Similarity is determined by finding the least difference between the features of one point or group of points and those of another. In this analysis, the minimum Euclidean distance between the centroids of the groups was used to select which groups to combine.

The dendrogram of the points using scaled, two principal components (Figure 8-2.) shows that five (intersections with the blue horizontal line) clusters probably provide the best groupings of the points. The Paleoindian composing each of those five clusters are shown in Table 8-1. The yellow highlighted projectile points are the proposed Sunrise points. Notice that Group 3 and Group 4,

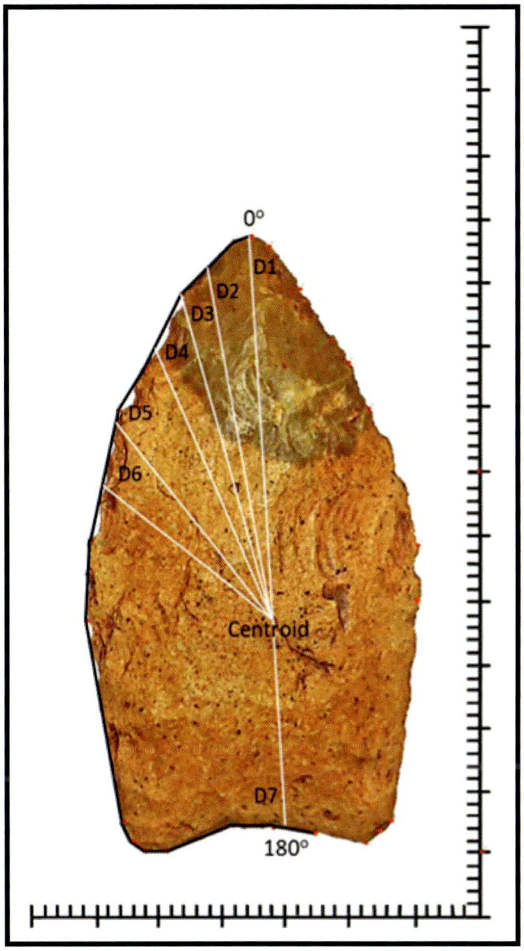

Figure 8-1. Describing the Projectile Point Using Lengths from the Centroid to the Edge.

which contain most of the Sunrise points, are more like Group 1 and Group 2 respectively, than they are to each other. They join those groups lower in the dendrogram than they join each other.

Figure 8-3. shows a plot of the two principal components. Each of the five groups is color-coded and an ellipse has been used to include as many of the points within each group as possible. The proposed Sunrise points are marked as blue boxes. The mean values of the points within each group are shown by the triangles. The closer a projectile point is to the triangle for that group, the more it will look like the mean shape for that group as shown in Figure 8-4. Conversely, the closer a projectile point is to the boundary between two groups the more it may resemble projectile points in that adjacent group. These are the projectile points that subjective grouping is most likely to place in another group.

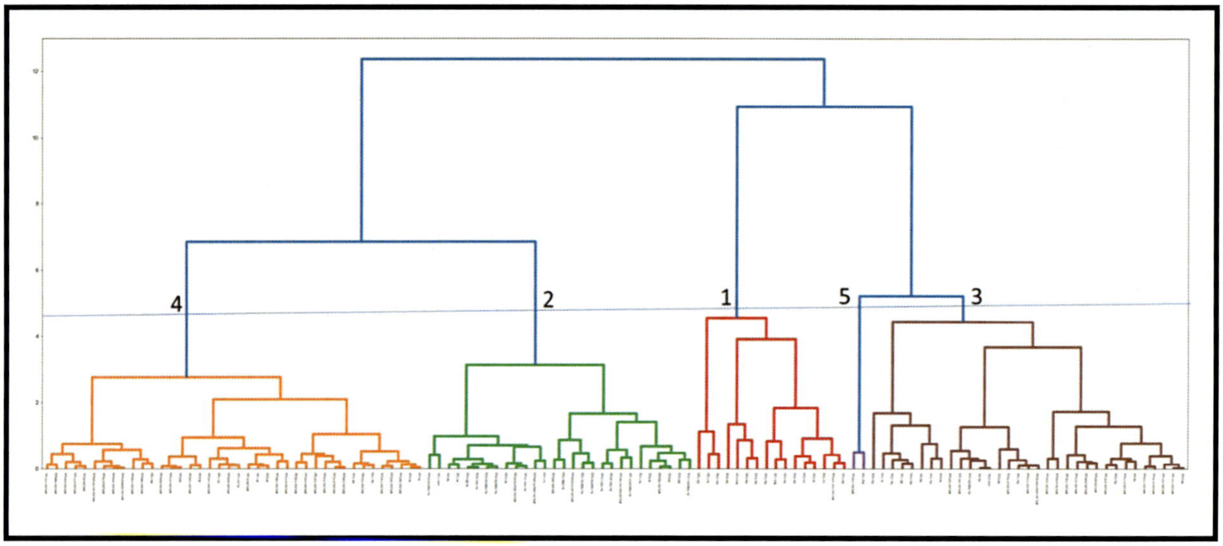

Figure 8-2. Dendrogram of Clustering of Principal Components with Five Groups Indicated.

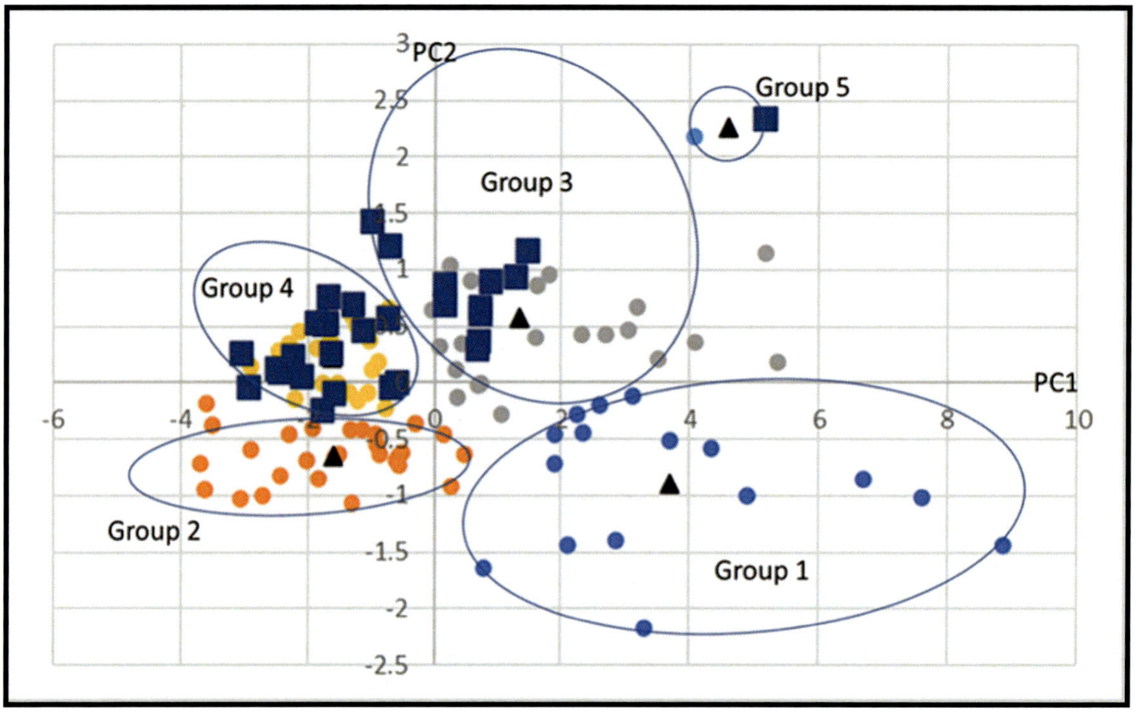

Figure 8-3. Plot of the Two Principal Components Showing the Five Groups, Their Mean Values and the Sunrise Points.

These are also the projectile points most likely to move to another group when points are added or subtracted from the dataset or other changes are made in the mathematical grouping approach.

It is troublesome, but not surprising, that other projectile points previously classified as other types of Paleoindian also occur in those groups. From Figure 8-3. it is apparent that the distance of many of those non-Sunrise points from the mean of the respective groups is no different than the proposed Sunrise points. This would imply that either the other points in a group now would be labeled as Sunrise points or breaking the dataset into more groups might separate the Sunrise points from the others (See Appendix 8-1.), or there are other attributes that distinguish them from the other Paleoindian points. From Appendix 8-1. it is obvious that selecting more groups, further down in the dendrogram, does not separate out the Sunrise points.

Since point size was not normalized across the dataset in the analysis above, it may be that size is influencing the cluster membership as much or more than some of the other parameters that describe the points. If the shape and size of the "mean projectile point," defined by the mean value of each of the 180 parameters for all points contained in a group (Figure 8-4.) are compared, it appears that size is a difference between Group 3 and Group 4. Both shape and size appear to be different between Group 5 and Group 3 and Group 4.

After examining the results of the analysis above, Bradley asked how the Sunrise points would cluster if the other Paleo-points were not included and therefore did not influence the mathematics of the clustering. This question was restated as: What if the proposed Sunrise points are assumed to be a single style, differing only in size? After correcting for size, how would the resulting groups compare in shape?

To investigate that question an average point was defined using the average distance from the centroid to the edge at each degree of arc using only the proposed Sunrise points. This set of average values was then assumed to be the average size and shape of the Sunrise point and used as the reference template.

Each proposed Sunrise point was then

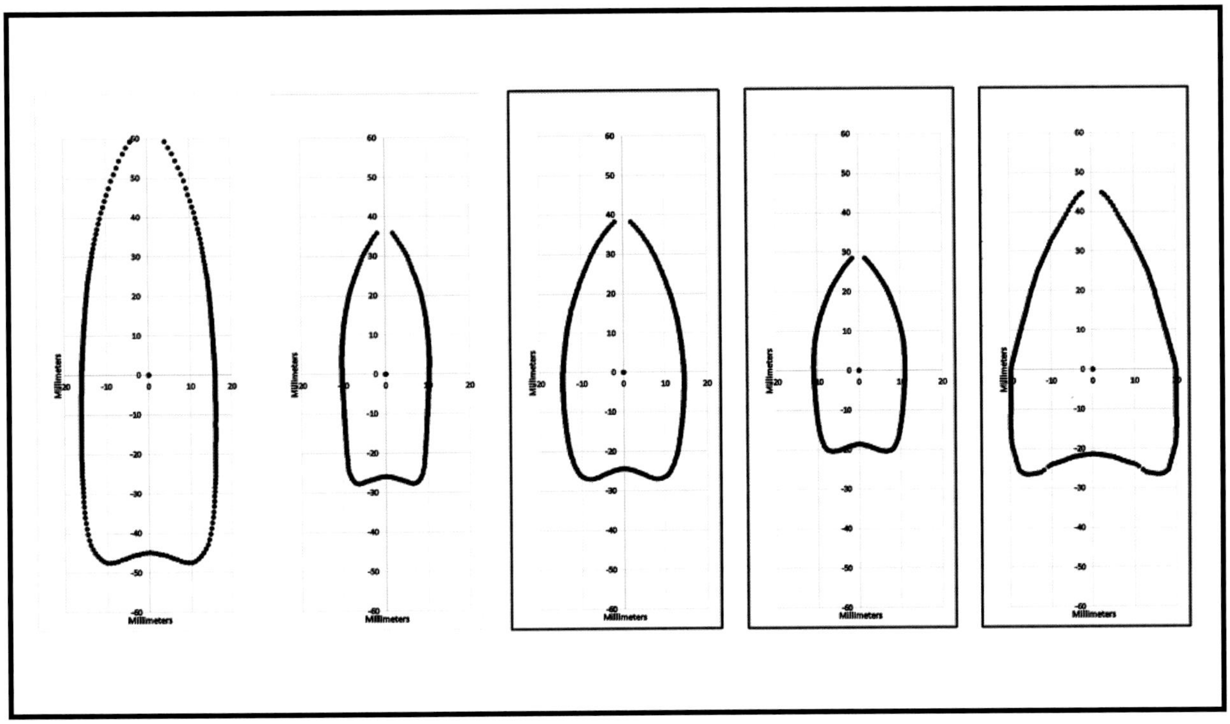

Figure 8-4. Mean Shapes and Sizes of Points for Five Groups Derived from Clustering Principal Components (three boxed groups contain proposed Sunrise points).

Group 1	Group 2	Group 3	Group 4	Group 5
48PL330-1124-1125.jpg	48PL330-1964.jpg	48PL330-1114.jpg	48PL330-1115.jpg	48PL330-1144.jpg
Bwd-1.jpg	48PL330-1965.jpg	48PL330-1116.jpg	48PL330-1118.jpg	slug-1.jpg
Bwd-2.jpg	48PL330-1968-FS1064.jpg	48PL330-1119.jpg	48PL330-1126.jpg	
Drk-1.jpg	48PL330-3915-FS408.jpg	48PL330-1123.jpg	48PL330-1134.jpg	
Drk-2.jpg	48PL330-3921-FS1165.jpg	48PL330-1131.jpg	48PL330-1135.jpg	
Drk-3.jpg	48PL330-3922-FS724.jpg	48PL330-1132.jpg	48PL330-1136.jpg	
Drk-4.jpg	Han.jpg	48PL330-1140.jpg	48PL330-1137.jpg	
Drk-5.jpg	Lnr-1.jpg	48PL330-1941.jpg	48PL330-1138.jpg	
Flc-1.jpg	MI-10.jpg	48PL330-1970.jpg	48PL330-1143.jpg	
Lnr-2.jpg	MI-11.jpg	48PL330-2169.jpg	48PL330-1146.jpg	
Lnr-3.jpg	MI-6.jpg	48PL330-2173.jpg	48PL330-1147.jpg	
MI-3.jpg	MI-7.jpg	48PL330-2178.jpg	48PL330-1158.jpg	
MS-4.jpg	MI-9.jpg	48PL330-2179.jpg	48PL330-1963.jpg	
Wyc-4.jpg	Mid-2.jpg	48PL330-3167.jpg	48PL330-1968.jpg	
Wyc-5.jpg	Plv-EB-4.jpg	48PL330-3388.jpg	48PL330-1969.jpg	
Wyc-6.jpg	Plv-ORM-7.jpg	48PL330-3936-FS641B.jpg	48PL330-1972.jpg	
	Plv-TMM725-0-1.jpg	48PL330-823.jpg	48PL330-1976.jpg	
	Plv-TMM725-1.jpg	Cby-1.jpg	48PL330-2000.jpg	
	Plv-TMM725-13.jpg	Cby-2.jpg	48PL330-2170.jpg	
	Plv-TMM725-23.jpg	Cby-3.jpg	48PL330-2172.jpg	
	Plv-TMM725-4.jpg	Cby-4.jpg	48PL330-2176.jpg	
	Plv-TMM725-8.jpg	Csp.jpg	48PL330-2177.jpg	
	Plv-TMM725-9.jpg	Dent.jpg	48PL330-2183.jpg	
	Plv-TWC-10.jpg	Jfc-1.jpg	48PL330-3180.jpg	
	Plv-TWC-13.jpg	MS-1.jpg	48PL330-3404.jpg	
	Plv-TWC-40.jpg	MS-2.jpg	48PL330-391-FS-908.jpg	
	Plv-TWC-9.jpg	MS-5.jpg	48PL330-3935FS74.jpg	
	UnkC-1.jpg	Plv-TMM725-2.jpg	48PL330-41.jpg	
		UnkC-2.jpg	48PL330-45.jpg	
		WS.jpg	Lap-1.jpg	
		Wyc-2.jpg	MI-1.jpg	
		Wyc-3.jpg	MI-2.jpg	
		Wyo.jpg	MI-4.jpg	
			MI-5.jpg	
			MI-8.jpg	
			Mid-1.jpg	
			MS-3.jpg	
			Pin.jpg	
			Plv-1.jpg	
			Plv-CS-1.jpg	

Table 8-1. Projectile Point Membership of Each Group Resulting from Clustering Principal Components

compared to the reference shape. This was done by determining an average size scaling factor between each point and the reference template and applying that factor to the value at each degree of arc for that point. This provided a best fit between each proposed Sunrise point and the reference template. This removed size differences between points and left only the shape differences. The best fit values at each degree of arc for each of the proposed Sunrise points were then used in Principal Components Analysis and five principal components resulted. A dendrogram was created using the resulting five principal components (Figure 8-5.).

The mean shape of the three groups of points (intercepts of the black horizontal line in Figure 8-5.) are shown in Figure 8-6. The points that are in each of the three groups are listed in Table 8-2. In Figure 8-6. there are visible differences between the three mean point shapes. All three have concave bases but have different shapes of the blades. However, it is easy to imagine a continuum of shapes that have been divided into these three groups based on the similarity of all 5 principal components. However, because of the small number of proposed Sunrise points used in the analysis these results should be considered with caution.

It should not be a surprise when results of GMM clustering do not agree with subjective grouping of points. GMM, as shown in this analysis, treats all inputs equally and provides a repeatable result when the same inputs and mathematical criteria are used. Does it reproduce the mental template of the knapper? Probably not. But it does provide insight into repeatable similarities and differences between points in space and time. Adding the results of this tool with attributes, such as the technology discussed in Chapter 7, may allow archaeologists to move closer to understanding the role(s) projectile points had for different groups of people and at different times.

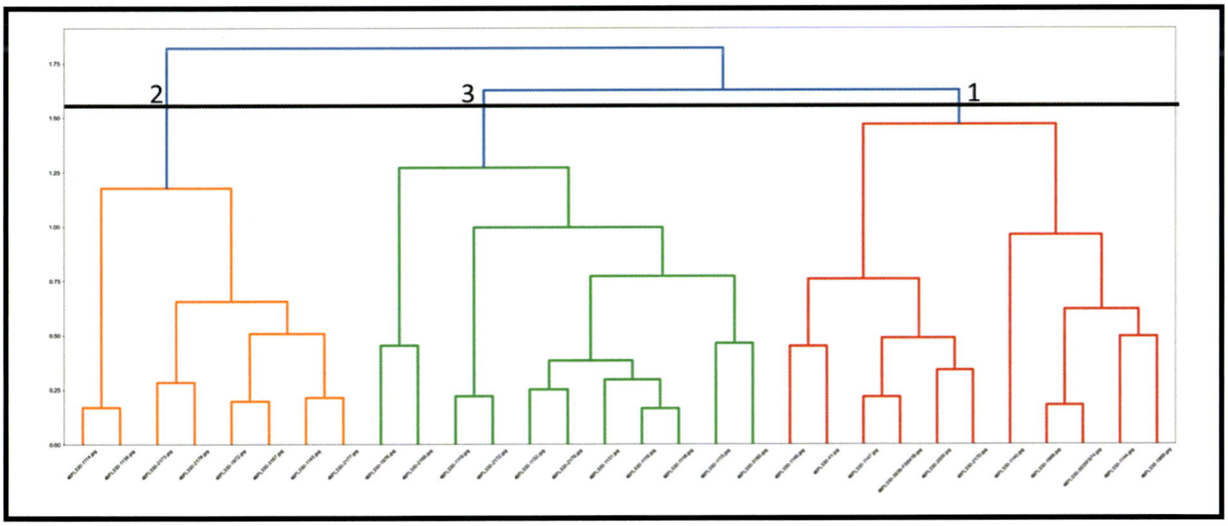

Figure 8-5. Dendrogram Using Shape of Sunrise Points Only.

Group 1	Group 2	Group 3
48PL330-1140.jpg	48PL330-1114.jpg	48PL330-1115.jpg
48PL330-1144.jpg	48PL330-1138.jpg	48PL330-1116.jpg
48PL330-1146.jpg	48PL330-1143.jpg	48PL330-1118.jpg
48PL330-1147.jpg	48PL330-1972.jpg	48PL330-1119.jpg
48PL330-1968.jpg	48PL330-2173.jpg	48PL330-1131.jpg
48PL330-1969.jpg	48PL330-2176.jpg	48PL330-1132.jpg
48PL330-2000.jpg	48PL330-2177.jpg	48PL330-1976.jpg
48PL330-2170.jpg	48PL330-3167.jpg	48PL330-2169.jpg
48PL330-3935FS74.jpg		48PL330-2172.jpg
48PL330-3936-FS641B.jpg		48PL330-2179.jpg
48PL330-41.jpg		48PL330-3180.jpg

Table 8-2. Mean Shapes of Sunrise Points in Three Groups Based on Shape

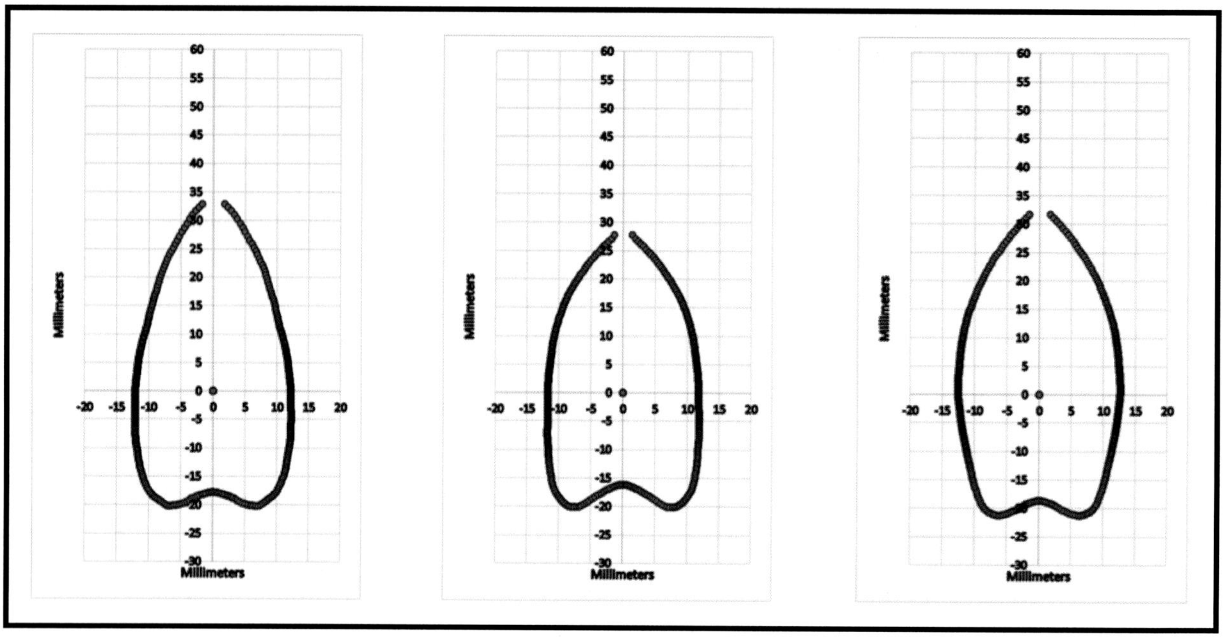

Figure 8-6. Mean Shapes of Sunrise Points in Three Groups on Shape.

Chapter 9

Chipped Stone and Osseous Tools

George M. Zeimens

Disclaimer: Several problems were encountered in attempting to identify the raw materials from which the lithic artifacts are made. First, after languishing for millenniums in the ochre mine deposits, the artifacts became so saturated with hematite that original colors were difficult to determine. Also, apparently some sort of chemical modification took place that made it difficult to ascertain original texture in many, if not most cases. That situation was further complicated by the fact that, the John Voight (JV) source chert and quartzite, of which the majority of the artifacts from the ochre mine are made, are naturally high in iron content, which further affects not only color and texture, but also weight. Of major consideration is that the range of variation in color and texture of the material in the JV source is phenomenal. Some of it is indistinguishable in color and texture from material found in other formations in the Hartville Uplift (HU), and some of it is dendritic which is one of the hallmarks of HU chert. And some of it overlaps visually in color and texture with Knife River Flint (KRF) and Alibates (AL).

Analysis using ultraviolet irradiation produced mixed results. Several artifacts from the talus and ochre mine fluoresced, but so did several KRF and AL look-alike samples from the JV source. The ultraviolet fluorescence test has been found to be an unreliable method for determining material type, at least in the case of KRF (Kristensen et al. 2018; Evilsizer 2016). However, several specimens from the talus and from the test in the ochre mine contain white splotches (internal mottling), or plant micro fossils characteristic of KRF, and are likely of that material type. In view of these problems, our analysis of the various materials should be considered to be tentative until such a time that additional methods can be applied. Currently, the lithic raw material in the Powars II collection does not serve well for considerations such as long-distance travel or cultural movements of Paleoindian people.

This treatise on the tool collection is intentionally descriptive, and is not meant to be a detailed analysis. The collection from the site is amazingly large, but as stated earlier, all tools were recovered from a mixed context where multiple cultural complexes are represented. Other than projectile points and some of the point preforms, none of the stone tools are assignable to any specific complex. Nor is it possible to determine how different tool types may have functioned together for specific activities. And another complicating factor in assigning the artifacts to specific tool categories, is that many of them are composite tools that exhibit use on multiple edges. For example, some exhibit scraping and cutting use on more than one edge, and may contain denticulate and graving tips as well. Yet, with more intensive scrutiny, the collection does have the potential for numerous in-depth analyses in the future. As we have suggested earlier, several lines of evidence may indicate that many of the tools were used elsewhere and deposited in the ochre mine for an undetermined reason. However, if the tools were used in the mine, determining how they may have functioned in a site as unique as the ochre mine will be challenging. Hopefully, analyses of the collection will be bolstered by more meaningful data, when stratified cultural components are examined during future research, in both the deposits adjacent to the ochre mine feature, and in the terraces on the valley floor below the site.

Spokeshaves

There are 37 specimens in this category, all from the mixed talus deposits. Twenty-five of them are made of material from the JV source, three are probably of KRF while nine are of the 'Classic' HU materials. They are generally made on flat flakes with sharp edges and are mostly somewhat longer than they are wide. They range from 29mm to 69mm in length with an average of 48mm. Width is highly variable. Twelve are made on flat blades. All of them are unifacial, and all are actually composite tools that exhibit use on edges in addition to the concave, spokeshave grooves in the form of prepared, straight and convex, beveled scraping edges. Six specimens contain sharp, pointed gravers either on a lateral edge or end. Two also once served as end scrapers, one with a spur on the working end, and the other has a highly worn and polished working end. Overall, there are two relatively uniform sizes of spokeshave grooves. Twenty-six are of a large size and average 22mm in width, and 6.5mm in depth at the center of the groove. Eleven specimens constitute the smaller size, averaging 8mm in width and 4mm in depth. Angles of the beveled edges in the interior of the grooves range from steep (65°) to very acute with variations in between. All of them are used artifacts in that, working edges show various degrees of use in the form of both crushing and polish. Four specimens are fragments for which we found no refits (Figure 9-1.).

Some of the spokeshaves exhibit grooves on opposite edges positioned at opposite ends of the tool (Figure 9-1., bottom row). They are made on blades and have a graver tip on one end. This unique type of tool has been found in Folsom components

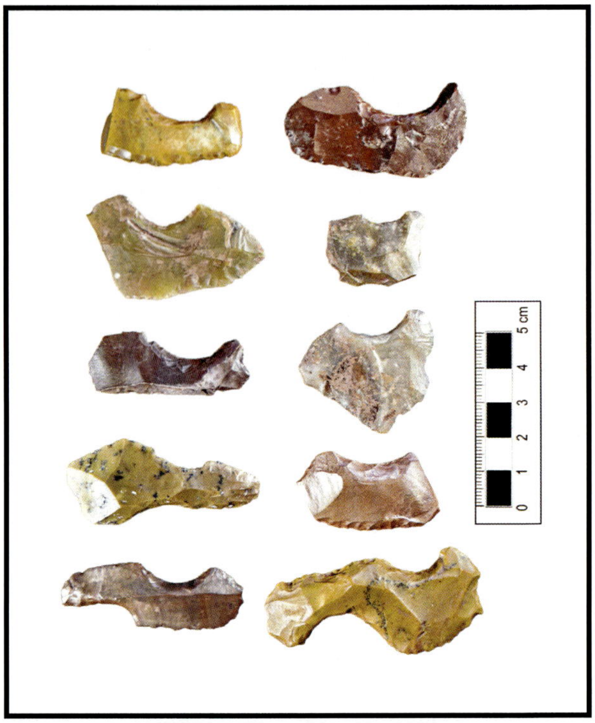

Figure 9-1. Spokeshaves

at other sites (Wilmsen et al. 1978; Hofman et al. 1998), and has not been found in other Paleoindian contexts that I am aware of. It is possible that in the future this artifact type may become a trait diagnostic of the Folsom cultural complex.

Gravers

The collection contains twenty-nine gravers. All of them are unifacial tools made on relatively thin flakes. Thirteen are single purpose tools with a single graver tip in the form of a sharp, prepared projection on one edge or end. Fifteen are multi-purpose, with one or more prepared and/or use-retouched scraping edges in addition to a graver tip. In some instances, a prepared graver tip is located along a beveled edge, suggesting that the tool had been used as a scraper prior to the addition of the graver tip. One specimen has graver tips on opposite lateral edges. Additionally, three spoke shaves also contain a graver tip, and two end scrapers have spurs on the working ends that are likely also gravers.

There are two groups of gravers based on size. Those in the group of smaller specimens are generally made on thin flakes, that range from 29mm to 38mm in diameter. They all have very fine, sharp needle-like tips. Those in the other group are generally thicker, heavier tools that vary from 45mm to 62mm in diameter, and their using tips tend to be thicker and heavier than those in the smaller group. Polish is easier to detect on the heavier tips. Nineteen of the gravers are made from JV chert, six from HU chert and quartzite, two from an unknown source, and one is made from hematite (Figure 9-2.).

Blades and Blade Tools

There are numerous ways that blades can be produced. They are often a fortuitous product, and use of them was apparently also sometimes opportunistic in that, when a sharp flake was needed they were picked up randomly, used momentarily, and then discarded. The original definition of a blade is that, it is a flake or tool that is twice as long as it is wide (Bordes 1961). If that definition were used here, there would be several hundred blades in the Powars II site collection, as well as in the collection from the JV lithic quarry. The definition used here defines blades, as a flake that is twice as long as it is wide and "....has subparallel blade facets on their exterior surfaces, is triangular, prismatic, trapezoidal or trapezium-like in cross section and have relatively even lateral margins" (Collins 1999 pp 7-10). This definition limits the site sample to a specific and easily identifiable type of flake better suited for studies in lithic technology (Bradley et al. 2010).

There are forty-nine blades in the Powars II collection. Eighteen of them are tools that exhibit prepared scraping or cutting edges, two have graver tips, three are spokeshaves, and one is an end scraper. Eleven show use on lateral edges in the form of crushing and polish. Twenty specimens show no use at all, and three of them appear to be crested blades. Thirty-five are made from JV chert, six are of HU chert, six are of KRF and two are of red Arikaree quartzite. Blade length is highly variable. The longest specimen exhibits no evidence of use and measures 112mm in length. The shortest is an end scraper 43mm in length (Figure 9-3.).

Miscellaneous Bifaces, Excluding Projectile Points and Projectile Point Preforms

The collection contains ninety-three bifaces. Only fourteen of them are whole or complete artifacts. Two of the complete specimens are small, very thin, ovoid in shape with percussion flaking across both faces. The edges around both pieces show heavy use in the form of use-retouch and polish. One is made of chalcedony and the other of chert, and both materials appear to be from the JV quarry. Another complete specimen is a thin,

Figure 9-2. Gravers

Figure 9-3. Blades and blade tools

triangular backed knife. In this case, the surfaces of both faces were thinned by the percussion method, and one lateral edge is ground relatively smooth, while the opposite edge is pressure flaked to an acute, sharp angle. The centers of both faces of the base were thinned by removal of flakes that extend up to near the center of the piece. That specimen is made of JV chert. Another complete biface is made of JV chert, and is so heavily laden with hematite that it could be mistaken for a chunk of iron ore. One end of that artifact is highly polished from heavy use. The remainder of the complete bifaces are elongated, relatively thick specimens that appear to be exhausted tools used for cutting and/or scraping, and are made of JV chert and red Arikaree quartzite.

Seventy-nine bifacial artifacts are aberrant fragments, many of which apparently were broken during an early or prefom stage of manufacture. The type of tool they were intended to become is not apparent. Thirty-four of the fragments are made of JV chert and quartzite, seventeen from HU chert and quartzite, twelve of red Arikaree quartzite, two of porcellanite, six of platey chert, while eight are of an undetermined source of quartzite. Given the large number of fragments, both from the talus and the test in the ochre mine, one would expect to find some refits, yet there are none. It stands to reason that if these tools were used and broken in the mine, or were manufactured in the mine, it would be possible to recover some refits. It could be that as with the projectile points, these tools were used elsewhere and later deposited in the ochre mine (Figure 9-4.).

Figure 9-4. Backed knife, ovid bifaces, exhausted bifaces

Side Scrapers

Thirty of the side scrapers in the collection exhibit beveled edges on one or more sides. Working edges vary from straight to slightly convex. They are all unifacial tools. Twenty-four are complete specimens and six are fragments. The angle of bevels vary considerably, and in all cases they are on the dorsal face of the artifact. All are made on flakes, and five fit in the blade category. Several show use-retouch along edges other than the prepared beveled edges. Shape also varies, but most are relatively large, elongated or slightly ovoid tools averaging 70mm in length, and 47mm in width. Working edges show various amounts of use in the form of crushing and polish. The material that one specimen is made of contains a large amount of hematite, that renders it unusually heavy for its size. The working, convex, beveled edge of that piece is highly polished, an indication that it may have been use as digging tool. Sixteen are made of JV chert or quartzite, eight are of HU chert or quartzite, four are of KRF, one is of red Arikaree quartzite, and the material for the one heavily laden with hematite was not determined (Figure 9-5.).

End Scrapers

Eleven tools are end scrapers. Four are small, triangular tools that average 35mm in length, and three are larger and average 60mm in length. All are unifacial, and in all instances the bulb of percussion is at the narrow or proximal end. One specimen has beveled, working areas at both ends. Six have beveled lateral edges that appear to have been used for scraping. The bulb of percussion or proximal end of three specimens has been thinned, an indication that they may be hafted tools. Four have spurs on the edge of their working ends. Moderate wear is evident on the working ends in the form of crushing and polish. The angle of the bevel of the workings ends varies considerably. Four are made of JV chert, one is of HU chert, another is from an unidentified type of quartzite, one is a cortical flake of petrified wood, and four are of KRF material (Figure 9-6.).

Projectile Points

The collections from the talus and the test in the ochre mine contain one hundred and eighty-eight projectile points. Eighty-three are complete or relatively complete specimens. Cultural complexes represented include Sunrise, perhaps Clovis, Goshen, Folsom, Midland, Plainview, Agate Basin, Hell Gap, Alberta and Frontier. The Sunrise point type consists of thirty-four specimens, or 66% of the point collection and is the dominant point type from the site. This significance of such a large sample is that it proved to be a sound basis from which Bradley, Nunn and Hauser (Chapters 7 and 8 this volume) were able to distinguish the Sunrise point from other point types. It should be noted here that Wayne Powars had collected 42 projectile points from the site, that he presented to Dennis Stanford at the Smithsonian (Stafford 2003). Those points were not available for inclusion in the Bradley, Nunn and Hauser analyses.

Projectile Point Preforms

Sixty-one bifaces are categorized as projectile preforms. Several stages of manufacture are represented. Most are complete specimens except for twelve that are partial tips and bases. Several of them were abandoned due to flaws in the material or knapping errors. However, many could have been progressed into finished projectile points. For most of them, it was not determined what cultural complex they represent, but several were assignable to specific complexes. Sixty% are made from JV chert or quartzite, 21% are of HU chert and quartzite, 3% are of KRF while the remainder are of unidentified chert and quartzite. As is the case for the projectile points, I defer here to the Bradley and Nunn chapter for analyses of the preforms.

Use-Retouched Flakes

Utilized flakes in general exhibit very thin, sharp edges. Utilization is manifest in the form of crushing or polish along unprepared edges. They vary considerably in size and shape. Most often edge damage is minimal, and it is a difficult call as to whether a flake is a tool. The problem is that the thin, delicate edges can be damaged in a number of ways, other than from use associated with human activity. Stafford (1990 39-50) encountered this problem during his analysis of the collection from Frison's 1986 test excavation in the ochre mine. Stafford performed several experiments using flakes for various tasks, such as scraping rocks to obtain red ochre, and by sliding flakes downslope. Neither of those experiments produced the kinds of edge damage found on numerous flakes in the collection. Damage from use results in uniform

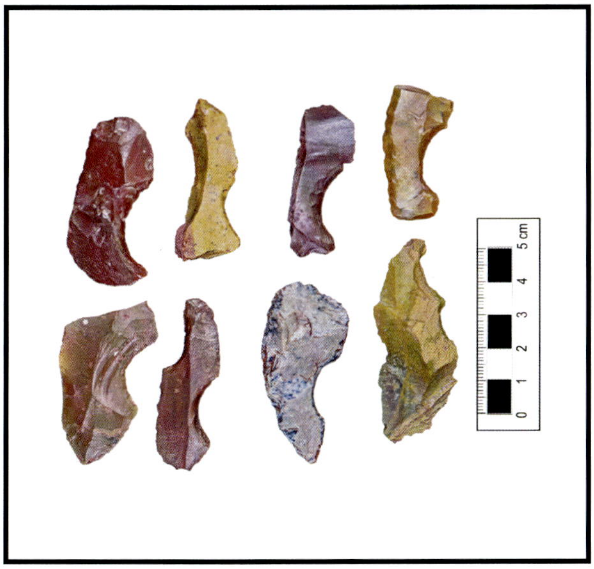

Figure 9-5. Side scrapers

Figure 9-6. End scrapers/spokeshaves

patterns of linear breakage along edges, rather than individual or random and uneven flakes (Ahler 1979 301-307). Polish is generally absent in the Powars II flake tools, except in rare cases where an edge was apparently used repeatedly on some hard surface.

Over six hundred flakes with damaged edges were examined. Only sixty-six of them qualified as tools. Various shapes and sizes are represented, and there seems to be no specific kind or size of flake that was preferred. The smallest, a blade fragment, is 24.5mm in length, 15mm wide, 7mm thick, and is trapezoidal. The largest is 65mm long, 45.5mm in width, 28.5mm thick and appears to be a core trimming flake. Eight specimens retain remnants of cortex on their dorsal surfaces. They are all unifacial, and quite often more than one edge has been utilized. Some tools appear to have been used more than others, but use is generally minimal. The sharp, thin edges would have been more suitable for cutting that for scraping. Fifty-six are made of JV chert, three are made of HU chert and quartzite, six are of KRF and one is of porcellanite. None of them exhibit polish that would have been produced from an extensive amount of scraping on rocks to obtain ochre. It seems likely that in most cases, use was an opportunistic event where a sharp tool was needed, was chosen at random, was used momentarily and then discarded.

However, it must be noted here that whether or not a flake qualified as a tool, was to a certain extent based on this investigator's personal judgment. There is no doubt that most of them are indeed tools, but it is possible that others should have qualified while some that did qualify are not tools.

Overshot Flakes

Two overshot flakes were recovered from the talus deposits. They measure from 37mm to 18mm across the distal ends. It was not determined if either of them were produced intentionally, and neither are diagnostic of any particular cultural complex. Both of them are made from JV chert.

Common Flakes

Salvage efforts in the talus produced over 7,000 flakes. They represent a variety of stages of tool production or maintenance. After languishing in the ochre-laden matrix for so long, the lithic material that they were made from could not always be determined with certainty. However, the material was dominated by JV chert and quartzite, followed by a small percentage of HU chert and quartzite.

Core Fragments

Thirty-nine core fragments were recovered from the talus and the ochre mine. They are all relatively small, multifaceted nodules, the largest of which measures 69mm x 56mm. Thirty-two are of JV chert, three are of JV quartzite and four are of HU chert. Three of the HU fragments retain small remnants of cortex.

Lithic Material

With the problems of raw material identification discussed above in mind, the material that the tools are made of include 78% JV chert and quartzite, 19% HU chert and quartzite, 6.9% KRF, 0.9% porcellanite, and 4.1% of materials not identified. These figures differ considerably from those previously reported (Frison et al. 2018, Pelton et al. 2022). The difference stems from the fact that, at the time of those reports, we were simply not yet totally familiar with the extreme range of variation in the material in the JV quarry, or how extensively the material had been modified visually from resting in the hematite deposits. These figures would also vary depending on the bias and experience of individual investigators. However, it is not likely that they would vary to any significant degree.

Determining the material that the projectile points are made of, was fraught with the same problems encountered with the tools. It is likely that no two investigators would agree on the outcome. With that being said, we determined that 65% of the points to be made of John Voight chert and quartzite, 8.5% from Hartville Uplift chert and quartzite, 7.5% from porcellanite, 3.3% from Knife River Flint, 3.7% from Arikaree Formation quartzite, 1.2% from Alibates, and the remainder from undetermined sources (Table 1.).

Examples of the non-Sunrise projectile points from the site are presented in Appendix 9-1. As with the Sunrise points, they are from mixed contexts and remain to be fully analyzed. It is interesting that almost every known Paleoindian cultural complex in the Western Plains region is strongly represented at Powars II except Clovis. A couple of the projectile points may be Clovis (see Chapters 7 and 8), but the classic Clovis blade core lithic manufacturing technology is not present. A similar situation exists at the Hell Gap site located just 4km or so north of Powars II (Larson et al. 2009), where numerous Paleoindian complexes are present but Clovis is missing. Bruce Bradley has identified one projectile point from Hell Gap as the Sunrise type (personal communication 2003). Again, it is critical that good, reliable dates be obtained for the Sunrise point, as it is possible that it is an early, contemporary or late Clovis variant.

Hammerstones

Relative to the number of other stone artifacts recovered from the talus and the test in the ochre mine, hammerstones were scarce. A hammerstone of the proper size, density and weight is essential for skillful knapping, although a suitable one is not always readily available. It is the type of tool that is likely to be transported from one activity area or site to another, and that may explain why so few were found in the talus, the ochre mine, or in the JV quarry (Bradley personal communication). Three of them, two from the talus and one from the ochre mine, have large flakes detached from the battered working ends, an indication of heavy use against hard objects. That kind of intensive use could not have been produced from breaking soft and crumbly paint iron nodules, that apparently were the subject of the prehistoric miners. For that reason, as is the case for the other stone tools, the hammerstones appear to have been used elsewhere before they were deposited in the ochre mine. That does not preclude the possibility that they were also used in the ochre mine (Figure 9-7.).

Red Ochre Balls

During the test excavation in the ochre mine, a ball, about the size of a common recreational softball, made of the highest quality red ochre was found, bagged and labeled. Unfortunately, that rare artifact never made it from the site to the lab, and is not available for further documentation. A similar ball was found during excavations in the JV toolstone quarry, and it measures 9.4cm in diameter and weighs 0.56kg. It is composed of compacted ochre mixed with small pieces of iron ore (Figure

Table 1. Projectile Point Lithic Material by Cultural Complex

Complex	Number of Specimens	Lithic Material
Sunrise	34	JV-23, HU-2, PO-3, KRF-1, AQ-2, UQ-3
Clovis	2	PO-1, UQ-1
Goshen	2	JV-2
Folsom	13	JV-9, HU-1, UQ-1, UC-2
Midland	9	JV-7, HU-1, UQ-1
Plainview	13	JV-6, HU-1, PO-3, AL-1, UQ-2
Agate Basin	21	JV-12, HU-2, KRF-2, PO-3, UQ-1, UC-1
Hell Gap	28	JV-24, HU-2, UQ-2
Alberta	1	JV-1
Frontier	16	JV-12, UQ-1, UC-3
Complex Unknown	48	JV-23, HU-7, KRF-3, AQ-5, PO-4, UQ-3, UC-3

(JV=John Voight chert & quartzite, HU=Hartville Uplift chert and Spanish Diggings Quartzite, KRF=Knife River Flint, PO=Porcellanite, Al=Alibates, UQ=unidentified quartzite, UC=unidentified chert.)

Figure 9-7. Hammerstones from the ochre mine that show heavy use from striking a solid object

Figure 9-8. Red ball of red ochre

9-8.). As was mentioned earlier, it is possible that tightly packed balls of ochre served to transport ochre from the mine to other sites.

It must also be noted here that George Frison had several tools from the site in his possession in his laboratory at the University of Wyoming at the time of his death. Those artifacts have not been located and were not available at the time of this writing.

Bone and Antler Tools

Very little bone material was present in either the talus or the test in the ochre mine. Most of the faunal assemblage is tentatively identified as bison with a few postcranial deer remains and several antlers (Pelton et al. 2022). Most of the specimens were poorly preserved and are very fragile and fragmentary. Several bones from the talus are obviously of historic vintage, in that they were better preserved, and had been butchered with iron tools. All of the antlers appear to be deer, and because they are in poor condition they contain little evidence for use, although we assume that they were used for digging. Most of the bone tools are fragments of long bones and ribs that are rounded and heavily polished on one end. Ten specimens are round, elongated fragments of bone rods, although it has not been determined if they are made of bone or antler. Three of the rods have round, smooth ends and the others are midsections. It is obvious that the shafts were polished, as polished areas are visible on the surfaces between splotches of deteriorated bone. One medial shaft is flat on one side and apparently was splintered, possibly from use. The precise diameter of the shafts is difficult to determine due to deterioration, but they vary little and average 15mm in diameter. Original length could not be determined from the fragments. The use of the rods is problematical, but it is possible that they are an element of weaponry (Frison et al. 2018, Figure 9-9.).

One inscribed bone fragment was recovered from the mixed deposits during the student test in the ochre mine (Pelton et al. 2022). It is a medial fragment from either a rib or dorsal spine, probably of bison. It is 132mm long and 34.5mm wide with a series of notches along both lateral edges. Very light scratch marks extend part way across the surface on both sides. It is possible that it is a portion of a bone hafting element (Figure 9-10.).

Figure 9-9. Bone tools

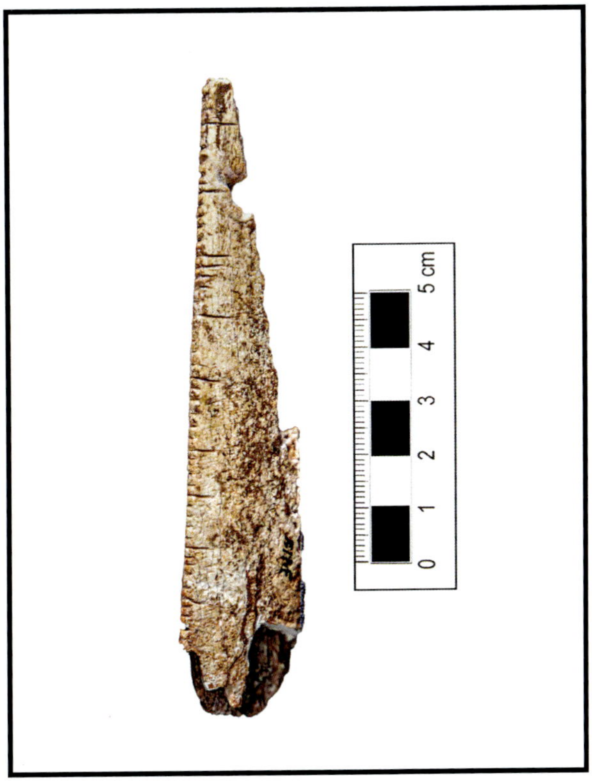

Figure 9-10. Inscribed bone fragment

Chapter 10

Erosion Abatement Efforts

Marcia Murdock,
Harold Hutson, and
George M. Zeimens

Natural erosion, along with disturbance from historic mining, combined to damage and threatened to obliterate the prehistoric Powars II ochre mine at Sunrise, Wyoming. The ochre outcrop is perched on a small bench of a spur along a ridge that joins a larger highland which was ultimately carved out into the historic Colorado Fuel and Iron mine. Development of access to the iron ore had carved off the end of the supporting spur ridge. Mining on the hillside created a huge pit known as the Glory Hole. Failing highwalls of the pit are located adjacent to the northeast corner of the ochre mine trench. A portion of the area of the ochre mine was further damaged by construction of a haul road along the hillside above the site. Dumping of mine spoils (waste rock materials) over the edge of the road buried (but actually protected) part of the site by covering it. The hanging wall of the cultural deposits was left exposed and highly vulnerable to erosion when the lower part of the hillside was bulldozed away to make room for several rail spurs. An undetermined amount of the site deposit was removed when the rail spurs were built.

As pointed out earlier, the ochre mine rests on a rather steep (roughly 20°) slope some twelve meters above the current level of the valley floor. The track hoe test discussed earlier (Chapter 3 this volume) determined that in the past the area immediately below the site was a minimum of 10m lower than it is today, and that the deepest part of the canyon extended along the north side of the valley. The present valley floor consists of leveled fill placed to facilitate construction of rail spurs leading to the Glory Hole and other mining facilities. Therefore, the deposits beneath the current valley floor in the area of the ochre mine contains historic fill (mine spoils) to an unknown depth.

The original extent of prehistoric cultural activity on the hillside is difficult to determine. Because of where it is situated, there is no doubt that the site deposits actually began eroding during the 3,000 year period that the ochre mine was active and erosion has continued up to the present time. One can only wonder how much of the site washed down the canyon over the years long before modern mining began. Based on the amount and rate of erosion that we can document during historic times, a large and significant amount of the site certainly vanished before historic occupation. Clearly a portion of the deposit below the site was later lost when the lower portion of the slope was bulldozed off creating a 12m vertical rock wall. After the canyon was filled with historic mine spoils, erosion continued and hundreds of artifacts slid down onto the historic level of the valley floor and into the talus below the site. It is obvious that during construction of the railroad just below the base of the rock wall beneath the site, much of the loose talus deposit was shoved around and an undetermined amount became mixed with ballast for the rail bed.

After a change in ownership, in 2014 we were granted access to the site and we were finally able to closely inspect and monitor the site deposits. We soon learned that after every rain, snow or big wind event, freshly eroded artifacts appeared around the edges of the site including projectile points, tools, flakes and bone fragments. In fact, most of the 1m x 2m test excavation that Frison had placed into the edge of the site deposits in 1986 had disappeared by 2014. Aside from weather, erosion was also exacerbated by deer regularly scrambling up and over the site to access a game trail that ran through the top of the site. Additionally, it appeared that some of the erosion resulted from surrounding mine reclamation done in the 1970s and 1980s by the Abandoned Mine Land program (AML) of the Department of Environmental Quality (DEQ) before it was known that the archeological site existed. Immediately after the site was identified, DEQ ceased AML reclamation activities in the area of the site. From what we could tell, AML reclamation did not encroach into the site deposits but did push material (old mine spoils and trash) over part of the site.

No doubt much of the site had been lost in the past from erosion so our immediate priority was to find a way to abate the erosion process. Putting some type of cover over the site appeared to be a good option for protecting the deposits from weather and game trails. In that regard, in 2016 we had hired a building contractor to design a building to serve as a cover. Engineering a building that high up on a steep slope presented a real challenge. What he proposed was to use pilings placed at ground level below the site that would extend up above the level of the site, and to build a cover attached to the pilings that would extend into the hillside above the site. The plan included a staircase attached to the pilings that zigzagged up to the level of the site. Construction of a road would be needed across the hillside for vehicle access. That plan would have protected the site. However, the design did not provide enough good access or working space for research crews or site visitors, and it was at best a short term solution to the problem since the structure would have to be continually disassembled and reassembled for access. To modify that design to provide for a more permanent and functional facility would have been extraordinarily expensive, and the cost for the project as proposed was far beyond our means. There was no doubt that this rare and unique Paleoindian red ochre mine was worthy of a large expenditure to stop the erosion, but we simply did not have the resources needed to find another solution.

We saw a news report that AML had funds for mine reclamation projects and thought that there might be a way to get a grant from AML. Geri Zeimens, a research volunteer, was finally able to contact Mrs. Marcia Murdock of AML to discuss the issue. Although AML does not provide grants for projects by other organizations, AML did have funds to conduct their own projects at abandoned mine sites. Already understanding the importance of the Powars II site, Marcia thought an AML project in the area of the ochre mine site might be possible if approached properly. She arranged to have Mr. Alan Edward, DEQ Administrator and other AML staff, meet with us at the site to review the situation and to determine if there was any action AML could take that would help. Marcia and Alan, along with representatives from the Office of Surface Mining Reclamation and Enforcement (OSMRE), AML's funding agency, and the State Historic Preservation Office (SHPO) traveled to the site to evaluate the situation.

Everyone at the meeting agreed that the valuable resources of the Powars II site were at risk of continued damage and deflation from erosion, and that considering the importance of the site, it needed protection. It was also determined that because of federal restrictions on how AML funds can be spent, site stabilization could be incorporated with a mine reclamation project that would alleviate abandoned mine hazards. AML and OSMRE agreed that as a mine that was at least 12,000 years old, and as the only known Paleoindian red ochre mine in North America, this site actually qualified as an abandoned mine. Therefore, the ochre mine's inherent hazards including the Archaic spoil pile and the modern highwall remnant both qualified as mine hazards in need of remediation. Site safety was a real concern due to the falling hazard posed by the highwall, as well as numerous large boulders in the mine spoil that hung loosely above the area where archeologists needed to work. The situation also posed a danger to visitors to the area. AML funding could be used for a hazard remediation project on the site. Edwards assigned Mrs. Murdock as the AML Project Manager and point of contact.

At the onsite meeting, various solutions to erosion abatement were discussed, including a possible retaining wall, some sort of cover, and finally Marcia proposed rebuilding the slope under the site to provide an earthen buttress to support the remaining site deposit. It was agreed that AML would pursue a project, and that AML would work closely with us and Mr. Voight to develop an acceptable stabilization plan.

Marcia brought her consulting engineer, Mr. Harold Hutson, PE, PG from BRS, Inc. to Sunrise to investigate the situation. Marcia, Harold, Frison and Zeimens met several times and considered various options for erosion control including discussion of the type of cover that the building contractor had designed. In the end, Marcia and Harold proposed a design that would place earthen material against the hanging wall of the deposit to rebuild the approximate original contour of the hillside. This would allow stabilization of the site and the redirection of water flows by building the slope up to the level of the site, thus covering the exposed edges and protecting them from the ravages of weather. The advantages to this approach were that it would eliminate the physical hazards to the ochre mine site, stabilize the remaining deposits, provide ample working surfaces for archeological crews, and provide a safe approach to the site from above. We all agreed that the plan was feasible and a good idea.

The final design employed Carlson Natural Regrade software to design a geomorphically stable slope with natural-looking drainages that would direct runoff around the site. Material excavated from the CF&I haul road above the site as well as material that was placed during 1986 reclamation was planned to be removed and used to construct the downslope fill. More backfill would be derived from nearby mine spoils. A long segment of chain link fence would be removed from across the site and later replaced with buck-and-rail fence. During the winter Marcia and Harold engineered the plan, gained funding approval, and contracted with various companies to do the work. Landowner John Voight concurred with the plan and generously agreed to allow use of additional material on his property to build the buttress. He also agreed to allow AML to use some of the topsoil he had stored nearby to help with revegetation. The project was to be closely monitored by Frison and Zeimens to assure that the site was not inadvertently damaged by earth moving machinery.

Work began on the project in the fall of 2019 and was completed later that year. The contractor was Earthwork Solutions of Gillette, Wyoming led by Wes Clason. The first stage of work used a special long-reach tracked excavator to salvage all the loose scree from the hanging wall and steep slope below the ochre mine site. This material was carefully reserved and stockpiled in a location that would not be impacted by construction so that it could be screened for artifacts. This proved to be very beneficial because it allowed retrieval and preservation of a number of artifacts that would have otherwise been lost.

Because the site hung some 12m above the valley floor at the top of a remnant highwall, twelve thousand cubic meters of material were needed to rebuild the hillside and apply an earthen buttress against the exposed edge of the ochre mine. This allowed the native slope of the site to transition smoothly into the valley floor and to eliminate the highwall hazard immediately below the site. Material was available from fill just above and adjacent to the site and from the berm that AML had placed in front of the Glory Hole. Removal of the mine spoil and road material uphill from the site deposits was carefully monitored by Zeimens and Earth Work Solutions personnel. Everyone involved worked as a team to insure the site was not harmed during the sensitive operation. At the hanging wall, geotextile fabric was placed against the wall face and raw edge of the eroding site deposit as a divider between the native ground with artifacts and the backfill. This was done to protect the artifacts and associated data and leave a demarcation between the cultural deposits and the new fill. This effort will also aid future investigators in determining the extents of the disturbed and undisturbed areas.

The grading design allowed the contractor to construct a surface which appears close to a natural slope and to complement the site contours. To protect the site from further erosion, this design carefully redirected runoff from the slopes above away from the known and potential cultural laden deposits. After the final grading was completed, the project was covered with topsoil and seeded. A buck rail fence was installed uphill from the site both to control access to the Glory Hole and to prevent livestock from wandering into the area.

The total cost of the construction project was $244,360.30. All construction and engineering costs were financed entirely with AML funds. We are now confident that erosion of the exposed edges of the site has been successfully controlled. Not to have attempted to take measures to prevent the loss of an archeological site as rare and unique as the Powars II Paleoindian red ochre mine would have been unconscionable. In addition to funding the work, AML funded the cost of preparing the nomination of the site for inclusion in the National Register of Historic Places, an important status for future preservation of the site. We are very grateful to AML staff for their insight, expertise and cooperation in making the project possible and a success (Figures 10-1., 10-2.)

There is however, another issue with erosion at the site that will need to be addressed. Because of the way the site is situated on the steep slope, the uppermost or top of the cultural bearing deposits are exposed at the surface. Weather events that produce runoff, and particularly wind, erode artifacts out of the top of the deposits. Of course, the trouble with eroded artifacts is that they no longer remain in primary context which limits their value for research purposes. In an attempt to abate this kind of erosion we covered the area with heavy tarps. The problem with that solution is that the tarps deteriorate from sunrays and weather and have to be replaced quite often. Just the act of installing and removing the tarps induces more erosion. As caretakers of the site, we are frustrated because we know that a site of this caliber is worthy of a higher level preservation effort. What is desperately needed is a permanent cover over this precious site. In that regard, as per his wishes, a memorial fund for George Frison has raised $8,000 to be used toward the cost of a cover. We do not have a fixed estimate for a cover at this writing, but are looking into various design options and costs. Adequate preservation measures that reflect the significance of Powars II are needed, and soon.

Figure 10-1. Site before erosion control project

Figure 10-2. Site after completion of erosion control project

Chapter 11

Stone Circle Site
48PL302

George M. Zeimens

A very intriguing stone circle site is located on the hill above the Powars II site. On a straight line, the circles are situated roughly 200m to the northeast of the ochre mine. The circles were recorded during the cultural resource study conducted for a reclamation project for the abandoned, historic Sunrise iron mine (Whissen 1995). Because of their close proximity to Powars II, it is possible that the circular stone features are related to activities in the ochre mine.

The site consists of two separate groups of circles, and they could easily and probably should be considered as two separate sites. One group of circles is located in a low valley on a gentle, almost flat slope in the west part of the area. The circles in that group are relatively small and average 4.5m to 5.5m in diameter, except for two large circles that measure up to 10m in diameter. The other group is located a short distance to the east, and is situated along a low ridge above the valley. These features are somewhat larger than those in the western group, averaging 6m to 7m in diameter, although they do vary considerably in size. The group on the ridge has been rather extensively disturbed by roads and some of those circles are not easily discernible. Another difference between the features is that, the stones used to construct the circles in the eastern group are significantly smaller than those used in the western group. The circles on the ridge appear to be very similar to other stone circles or tipi rings common in the area. In contrast, the western group are unusual in that they are made of much larger and heavier stones spaced very tightly together. Some of those stones may weigh as much as 40 kilograms (Figure 11-1.).

In the past both sites have been a favorite area for artifact collectors. Consequently, temporally diagnostic artifacts are scarce in the vicinity of the circles. During our survey of the area, only a few Late Archaic and Late Prehistoric projectile points and other tools were recovered from among the circles on the ridge, and none from among the western group. Recently, we did find the base of a projectile point preform on the surface next to one of the large circles in the western group. The preform, made of knife River Flint, is similar in shape and form to those at the Powars II ochre mine.

We wanted to learn if there is a relationship between the circles and the Powars II site. As part of the Powars II site project, we decided to place a test excavation in one of the small circles in the westernmost group. That circle, numbered SCS-1, is located near one of the large circles. It is not perfectly round, and measures from 4m to 4.5m in diameter.

We placed a test unit in the interior of the feature that included the entire eastern 1/2 of the circle. We began working by hand in arbitrary 5cm levels, and immediately encountered a hard-packed floor from 3cm to 5cm below the surface. The level

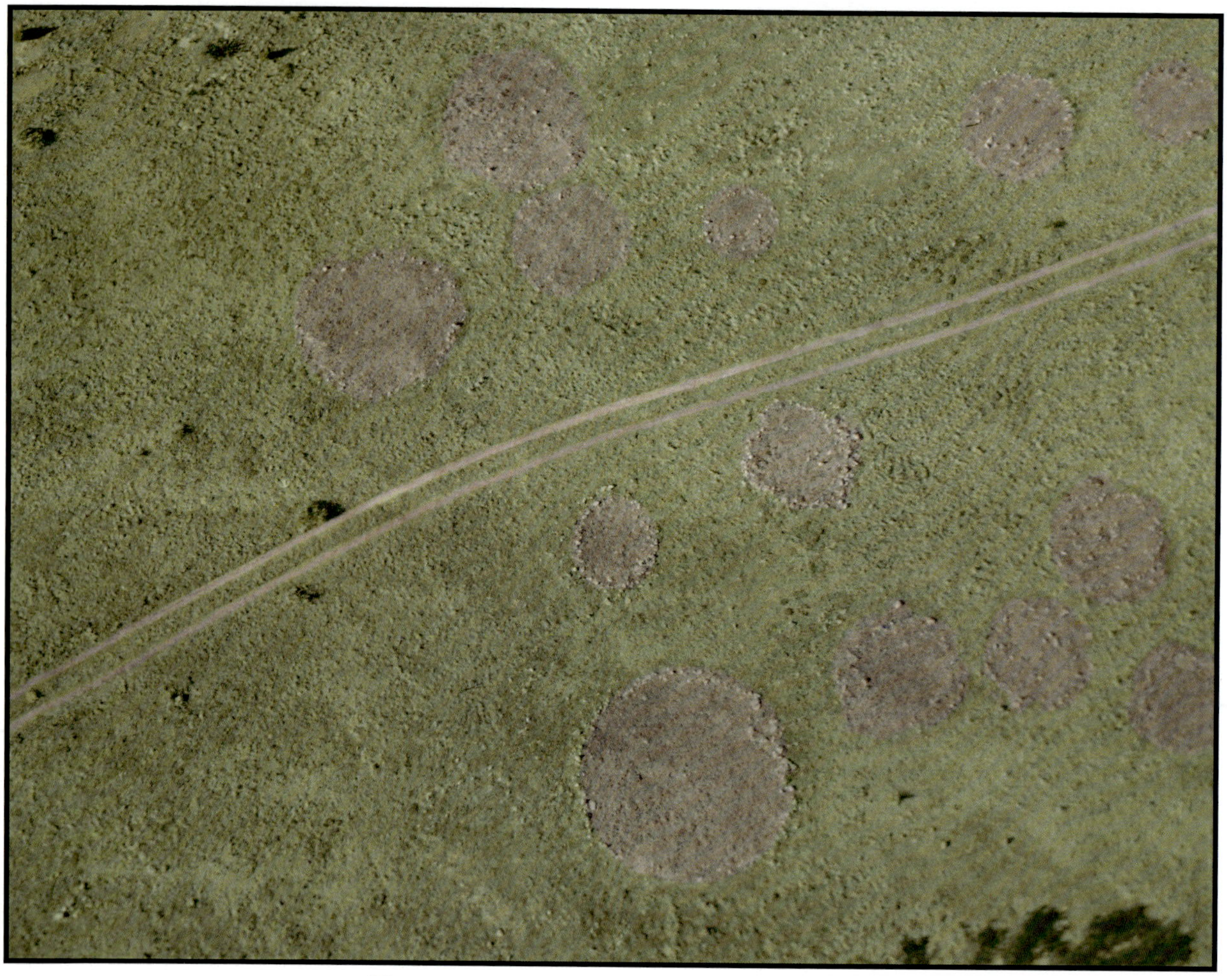

Figure 11-1. Drone photo of stone circles

of the floor is at the same depth as the bottom of the stones that comprise the circle. We found a few flakes, but because they were located so close to the surface, we could not determine perforce if they were associated with the floor. Numerous flakes are scattered across the surface in the area of the circles so the flakes could easily be intrusive. No bones or fire fractured rocks were present in the interior or exterior of the feature. When the floor was scraped with a trowel, a red streak was produced due to the fact that the floor is permeated with small fragments of pulverized soft or "paint" iron ore. There were no hammer or anvil stones present that might have been used to pulverize the ore. To piece-plot each small fragment of iron was not practical, as they were so numerous that individual plots soon became a solid blotch on the unit map. We discovered a pit on the floor in the center of the circle that measured 40cm x 42cm in diameter, and 10cm in depth. The pit walls slope slightly inward and the bottom of the pit is flat. The sides and bottom are stained a bright red. The soil that filled the pit was also red, and also contained numerous small pieces of highly fractured paint ore. There was no evidence for domestic activities such as cooking, hide working or flint knapping present in the pit or on the floor of the stone circle. There was no evidence to suggest that the stone circle might have been enclosed or covered, although it may have been.

It appeared that the primary activity that took place in the circle was processing of paint ore to produce red ochre. The small pit in the floor was certainly involved in the process, but just how it functioned is unclear. It is possible that water was somehow involved in leaching the ochre from the iron. For example, when it rains puddles of water on the valley floor below the ochre mine turn red with ochre. Because it is heavier than water the ochre settles to the bottom, and when the puddle dries the bottom is lined with pure, pasty red ochre.

However, more information is needed before the idea that water was somehow involved in processing the ore to obtain the red pigment.

The test had not resolved the question with regard to a possible relationship between the ochre mine and the circles. Although processing ore seems to be the best explanation for what had taken place on the floor of the circle, there was no direct evidence for the source of the ore. It could have been obtained from sources other than the Powars II mine, as there are numerous locations in the area where red ochre can be obtained.

To continue the investigation, we decided to test one of the large circles adjacent to SCS-1. Ten, 1m x 1m test units were lined out through the center of the circle designated SCS-2. The results of that test were almost identical to the test in the small circle. We found a shallow, hard packed floor near the surface that contained a few flakes and one spurred end scraper. Again, because the floor is situated so close to the surface, we could not determine whether the artifacts were the same age as the floor. The floor was riddled with very small pierces of pulverized paint ore and, similarly, there was no evidence for domestic activity. In this case, a few pieces of the ore were as large as 5cm in diameter, but most were much smaller. We encountered a somewhat larger pit, similar to the pit in SCS-1, situated near the center of the floor. That feature measured 75cm in diameter and is 15cm deep. The evidence suggests that the large circle functioned for the same activity as did the small circle. Because it is so large, 10m in diameter, it is tempting to assume that the purpose of the large circle included ritualistic activity, but we found no tangible evidence that it did. That being said, it is obvious that the main activity represented by findings on the floor of the circle involved red ochre, and red ochre is so often related to spiritualism (Morrow 2016; Roper 1991). It has been suggested that the large circle served as a dancing ring. However, the broken pieces of iron that covered the floor are so sharp, that they would likely have cut and severely damaged the dancers' feet and footgear (Figure 11-2.).

We excavated down to 50cm below the level of the floors of both circles, and found no evidence for earlier occupation of the site. There is one line of evidence that indicates that the circles are more recent than the Paleoindian ochre mine. A large percentage of the toolstone that the artifacts from the ochre mine are made of, was derived from the John Voight quarry or similar deposits. Most of the flakes scattered in the area of the stone circles consists of material from other sources. As stated above, there is no sure way to determine whether the flakes scattered on the surface of in the area are associated with the circles, or that the circles are related to the ochre mine. We did obtain a very small bit of charcoal from near the surface of the pit feature in the floor of the small circle, but it was determined to be too modern to produce a reliable radiocarbon date. Because the level of the floors is so shallow, charcoal collected for dating could easily be contaminated by roots and animal feces.

There are two sources of toolstone in the vicinity of the circles. One is located immediately south of the circles in a shallow drainage, and the other is on a low ridge on the north edge of the circles. Both are manifested by scattered flakes and core fragments, material that apparently was simply collected from the surface, as neither contain pits or other evidence of digging. The toolstone in the south source is a very distinctive, rather grainy quartzite, gray and slightly greenish and purple in color. One scraper and a few flakes of this type of quartzite were present in the Powars II ochre mine. The material on the ridge to the north is chert that varies in color, but is mainly a light tan or off-white or light red. Both sources appear to be related to an outcrop of dolomite of undetermined age.

Large stone circles such as those in the western portion of the site are fairly common on the Western and Northern High Plains, and it is often suspected that they are associated with ritual activities (Eddy 1977; Wilson 1981). In this case, the evidence indicates that they served for processing ore, and the idea that ritualism was a part of that process does not rise above the level of speculation. The Powars II ochre mine provides rare evidence for how the ore was mined, and the stone circle site provides evidence for how the ore may have been processed to extract the ochre. In Chapter 9, this volume, we reported that a ball of red ochre was found in the ochre mine trench, and that another was found in the John Voight quarry. It is tempting to speculate that tightly packed balls of paint ore served as a means of transporting the ore from the source to a processing area.

Figure 11-2. Test excavation in large stone circle SCS-2

Final Comments & Looking Forward

To say that the Powars II Paleoindian Red Ochre Mine is an incredible site is almost an understatement. When one considers the number of artifacts that have been collected so far, the number that surely remain in the deposits combined with those that must have eroded out of the site and washed downstream prior to the area being buried by historic mining, the situation is almost unfathomable. That does not include artifacts that surely exist in camps in the terrace below the ochre mine. We have yet to determine the range of activities represented by the wide variety of artifacts, other than mining for red ochre.

Several significant events, as presented in the various chapters in this book, contributed to securing the site and elevating it to a position of international interest. First, the property changed ownership and the new owner was approachable. Once the site became accessible, the community of artifact collectors rallied and came forth with funding to secure the site and to initiate research. Their support continues today. The aberrant projectile point that we have defined as the Sunrise point type, will be crucial to future studies of Early Paleoindian complexes. The large number of artifacts retrieved from the eroded deposits in the talus elevated the significance of the site, and brought more support. Work in the terrace below the ochre mine established the fact that stratified cultural remains exist therein. Artifacts in the terrace are present in deposits that date to over 14,000 years ago. The new type of toolstone discovered in the terrace is a significant find, and will have a profound effect on lithic studies at sites in the area far into the future. Also, our efforts to abate site erosion were bolstered by the Wyoming State Department of Environmental Equality, when they removed materials that they had left on the hillside during reclamation work in the 1980s. Those materials were a safety hazard to researchers working in the area. The agency then used the materials to build up a monumental berm unlike anything previously attempted at an archeological site, to protect the eroding edge of the site deposits and at no cost to us.

A nonprofit organization, the Sunrise Historic and Prehistoric Preservation Society (SHAPPS), has been formed to manage the site and the site collection. The collection will remain in a vault in the historic YMCA building at the site, and will be available for future research in the area. The CF&I Sunrise iron mine, including the town of Sunrise, is listed on the National Register of Historic Places as site number 48PL1391. The Powars II red ochre mine is also listed on the National Register, site number 48PL330. The John Voight toolstone quarry, and the terrace in the valley below the ochre mine, are not yet on the Register, and have state site number 48PL2366. The ochre mine, the toolstone quarry, and the terrace are located within the boundary of the historic iron mine site. The sites are all on private land owned by John Voight. SHAPPS is working with John to acquire the portion of his property that includes the sites in order to secure them permanently. To do so will require a substantial financial investment. Given the cultural significance of the property, SHAPPS is hopeful that the entire site complex can be acquired in the near future.

Dr. Bruce Bradley joined me as co-Principal Investigators at the site after Dr. George Frison passed. Having Bradley involved would have pleased Frison, and I could not be happier about it myself. Dr. Steven Howard, with Eastern Wyoming College, has joined our team and will offer a summer field school and other educational activities at the site. Since it has been determined that the deposits in the ochre mine feature are culturally mixed, future research should focus on the deposits in the terrace. The terrace is a vast area with room for many research projects for many years to come.

George M. Zeimens

"I think you should invite Doc."
A Final Comment by Elmer Guerri

"I think you should invite 'Doc'", Dennis suggested. "He's (Dr. George Frison) excited about a site he believes is one of the most important sites in North America. I think he would like to tell your group about it."

A group of eager enthusiastic artifact collectors was planning an exciting weekend with Dennis Stanford and Pegi Jodry in their lab at the Smithsonian Institution, a "Mecca of Paleoindian archeology", for those planning to make the trip. Members of the group weren't interested in choosing sides in the ongoing debate about "Peopling of the Americas". To a number they were interested in hearing about some of the latest research and developments in a field of science where even more promising revelations appeared to be on the horizon. Each of them, in their own small way, was interested in contributing to and being a part of that excitement.

Every one of those knowledge seekers had at one point or another, through their efforts and through their financial contributions, helped to support research of some of America's leading collector-friendly archeologists, anthropologists, and paleoarcheologists. Folks who were a part of the group included James Cox, Deanna and Elmer Guerri, Jim Lamson, Sherry Link, Sandi and Mike McGonigal, Rick Miller, Steve and Louisa Vaughn, Carl Yahnig, and Susan and Jerry Zanandrea. Special guests of the group included, in addition to Dr. George Frison, his daughter Carol Placek, and Geri and George Zeimens.

The February 15, 2014, group did not understand at the time of their visit that what transpired that exciting and inspirational weekend in the form of their encouragement and their initial generous financial contributions, would become the embryo that would eventually morph into the successful archeological effort out of which would evolve this most significant and commanding book. This illustrates how a small dedicated group could reach into their own magical bottle and bring forth an important vivid red pigment which would favorably paint an entire cooperative group with the color so prevalent in that magical location called Powars II.

We're all very happy that we did indeed invite 'Doc'. What grew out of that initial spore for all who are humble enough before science, is the understanding of the value of working together toward new discoveries, new learning, so that all may better understand who, when, why, how, and whence early humans came to this wonderful blessed land that is like no other on the face of the earth. The science behind "Peopling of the Americas" was greatly enhanced because a small group of artifact collectors thought it was important to have invited 'Doc'.

Research at the Powars II site has been funded entirely by several members of the Smithsonian tour group and other private parties. Support for the site began in 2014 and continues today. None of them are "pot hunters" which is true of the majority of artifact collectors. They all have a deep interest and appreciation for the craftsmanship and beauty of artifacts, the mystic of how they were made, and how they serve in understanding prehistoric life ways. They also enjoy reading about what has been learned from professional investigation at sites. Most of them regularly contribute to research by volunteering at sites, by financial support and by subscribing to professional journals. George Frison made no qualms about the fact that he began his career as an artifact collector. He stated that a lot of his career was bolstered by the fact that many of the sites he investigated had been discovered and reported by collectors and avocationals. Some of the major Paleoindian sites that fall into that category include Colby, Hanson, Agate Basin, Horner and Casper as well as many Archaic and Late Prehistoric period sites.

Elmer is a collector and recruits others to support research at Powars II and other sites.

Acknowledgements

Stabilization and securing the site would not have been possible without the cooperation and involvement of the landowner, Mr. John Voight (not the actor). John not only allowed access, but he contributed time and equipment for rearranging back dirt piles, backfilling, water screening, and made the crews comfortable while camping and exploring surface and subsurface areas of his property. He also not only allowed us to modify the landscape of a significant area of his property to abate erosion of the site, he contributed to that project by supplying material for that effort. Just imagine what condition the site would be in today had he adopted the 'hands off' attitude of the previous landowner. John also supported research by assisting in creation of the Sunrise Historic And Prehistoric Preservation Society (SHAPPS), for preservation and management of the site. He donated the historic YMCA building to SHAPPS for use as a museum and research facility. He donated the artifacts from the site to SHAPPS to be stored in the YMCA vault, where they will be available for future research. He has also been very helpful in numerous fundraising events that involved hundreds of people invading his property. He spends many hours mowing and grooming the area of the site in preparation for fundraising events. That the Powars II red ochre mine is now safe and secure, and the stage is set for future educational and scientific endeavors, is largely due to John's insightfulness and generous ways.

Marcia Murdock was inordinately instrumental in convincing the agency that she worked for (DEQ) that safety and erosion issues created by the agency needed to be resolved. It was her idea to address the problem of site erosion by shoring up the edges of the exposed site deposits by preparing a large berm that extended up the hillside. She arranged funding for the erosion abatement project, and for the cost of preparing the nomination of Powars II to National Register of Historic Places.

Research has been, and continues to be funded by the generosity of a consortium of artifact collectors and archeology aficionados. When they learned of the site, they immediately recognized its significance, and that with their support they could contribute to the archeological record in a meaningful way. They value what is learned from archeological scientific investigations, and recognize that knowledge serves to enhance the respect and admiration they have for the artifacts in their collections. It is their enthusiasm that keeps us prodding on in the arena of long days, hot sun, bugs, sore muscles and sweaty red-stained bodies. Our only pay is the excitement in their eyes and on their faces, when they see and touch an exquisite artifact that no one else has seen or touched for thousands of years. We recognize and stress the fact that, what can be learned from scientific archeological endeavors is not only the sole dominion of the archeological community, it is a part everyone's heritage.

A great deal of the progress achieved in developing and securing the site for the future can be attributed to creation of the SHAPPS organization. SHAPPS is a viable and dynamic outfit dedicated to preservation of the archeological and historical sources in Sunrise Park and the surrounding area. SHAPPS simply could not function without the support of the many, many financial contributors and volunteers that have donated so much of their talents, time, energy and equipment for the cause. There are far too many of them to attempt to recognize them all here lest we inadvertently miss someone, but we are very grateful to all of them. We are also greatly appreciative of numerous folks that helped with suggestions for, and editing of this volume, especially Garry Alkire.

Chapter 7

Appendix 1

Appendix 7-1. Powars II Biface Analysis.

Cat #	Phase	Material	Face A	Face B	Base A	Base B	Abandon	Max L	Max W	Th at Max W	Max W/Th	Weight	Comments	Affiliation
1007	lt	JV opl	ost	ovl	inv	inv	size	50	26.1	5.3	4.9	10.4	overshot, broken biface blank	Clovis
1017	lt	JV cht?	ovl	ovl	lateral	latt	unknown	51	26.6	6.3	4.2	10.5	square base	Sunrise?
2164	lt	SSt	ovl	flf	end-thn	eth	unknown	54	28.2	6.1	4.6	13	end-thn partly removed by lateral flaking	Sunrise?
1099	md	jsp	ovl	ovl	end-thn	abt	unknown	59	35	11.1	3.2	26.1	reworked distal fragment, unknown abandonment	unknown
3168	lt	SSt	prs thn	prs thn	prs thn	prs thn	too thick?	59	34.7	8.8	3.9	19.2	wide-spaced, deep prs flaking, some serial tip to base	Goshen or Frontier Complex?
1003	Ea/md	JV cht	ovl	abt mar	lateral	eth	unknown	60	35	6.9	5.1	17.1	abandonment?	unknown
3388	lt	JV agt	ovl	ovl	end-thn (3)	eth (1)	unknown	60	32.1	6.2	5.2	17.6	ovl followed by non-thn. Flat section, knob at tip. Abandonment?	Sunrise?
2165	Ea/lt	SSt	ovl	ovl	inv	crt	unknown	61	37.3	6.2	6	17.3	well made	unknown
1014	lt	jsp	ovl	ovl	inv ret	inv ret	Small, thick?	65	30	7.1	4.2	18.3	ovl, square tip Flake blank platform remnant? one mar hinge fracture, abandonment?	unknown
2182	Ea/lt	cht	non-thn	unm	abt	abt	unknown	66	44	10.5	4.2	35.1	cortical flake blank	unknown
1002	Lt/Md	JV agt	ovl	ovl	square	sqr	unknown	67	34.6	8.1	4.3	22.7	distal end-thn, curved	unknown
1019	Md	jsp	eth	eth	Eth (1)	Eth (2)	unknown	67	43.8	8.4	5.2	34.9	few flake scars, remnant 'square edge' possibly overshot remnant, end-thn both faces	Clovis
1006	Md	jsp	ovl	flf	ost	eth	unknown	68	43.5	8.8	4.9	34.8	cortical tip, Face A basal end-thn, Face B basal overshot remnant?	Clovis?
1207	lt	chl	ovl	com	inv non-thn	inv non-thn	unknown	69	39	9	4.3	30	regular, easy to finish into Clovis-like point	Sunrise?
1117	Md	cht	ovl	ovl	eth	lat	unknown	70	34.9	7.5	5.3	24.6	Face A basal end-thn, Face B double-stuck flake.	unknown
645	lt	SSt	ovl	inv mar	lat	inv non-thn	unknown	70	36.7	9	4.1	27.2	regular, easy to finish into Clovis-like point	unknown
3302	Lt/md	JV agt	non-thn	div	non-thn	lat	unknown	75	47.2	10.1	4.7	45.4	plano-convex cross-section	unknown
114	Ea	cht	non-thn	non-thn	lat	lat	stone, thick	82	40.2	16	2.5	40.6	poor material, hammerstone flaking	unknown
5618	Lt	SS	ovl, eth	div	end-thn (3)	eth (1)	unknown	85	45.8	8.2	5.6	43.5	eth on one face and diving flaking on the other	Sunrise?

Codes:

Phase- Ea= early; Md= middle; Lt= late

Material- sst= silicified sandstone; jsp= jasper; cht= chert; chl= chalcedony; opl= opalite; agt= agate

Source- JV= John Voight

Flaking- ost= overshot; ovl= overlap; prs= pressure; thn= thinning; unm= unmodified; eth= end thinning; non= non; inv= invasive; mar= marginal; abt= abrupt; com= comidial; div= diving; flf= full face; lat= lateral; ret= retouch; crt= cortical; sqr= square

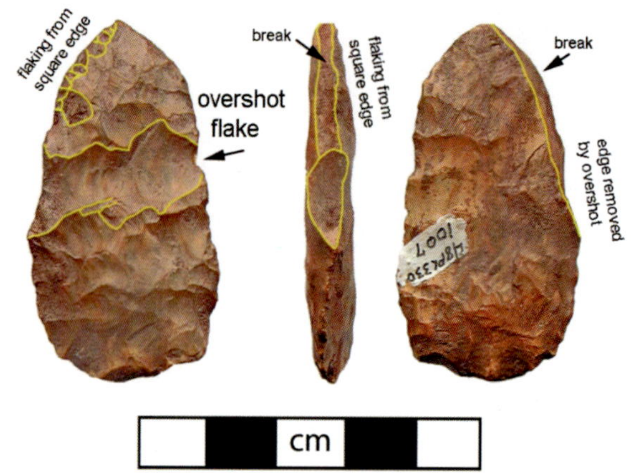

Biface 1007

Biface 2164

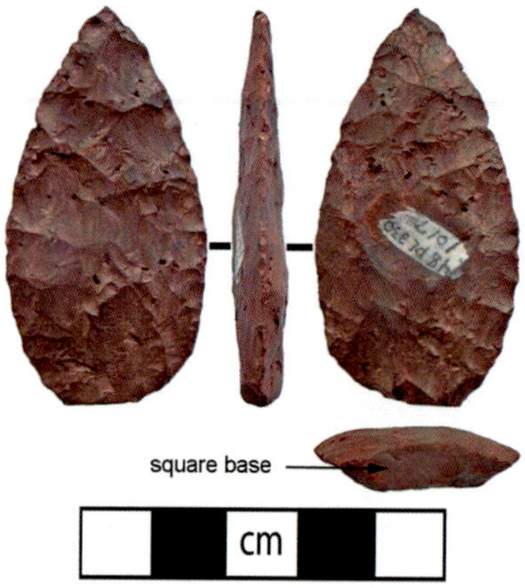

Biface 1017

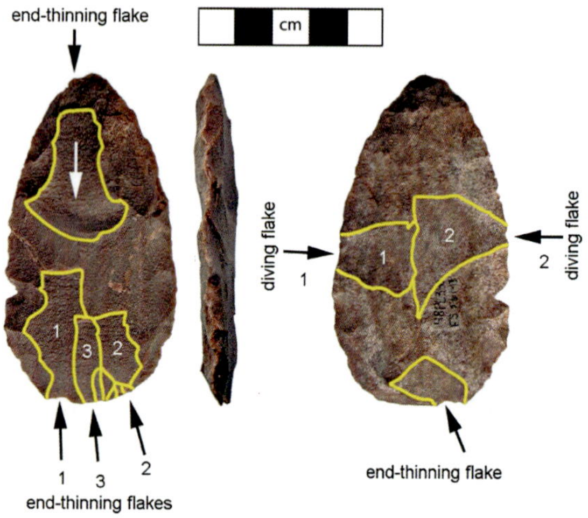

Biface 5618

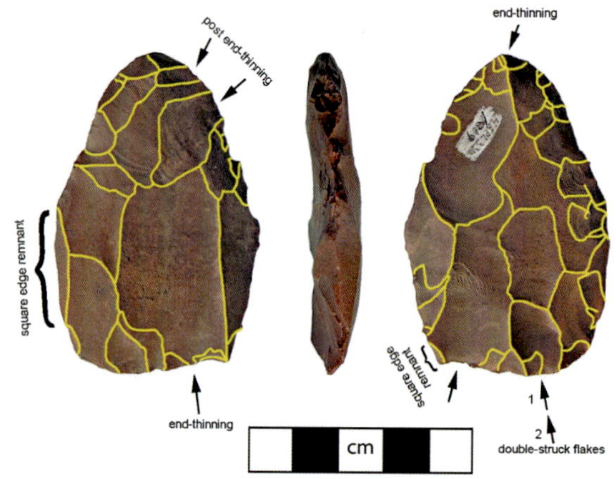

Biface 1019

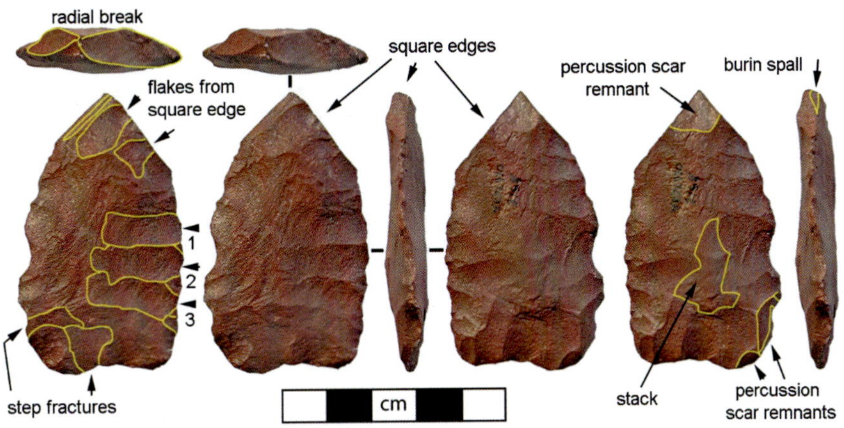

Biface 3168

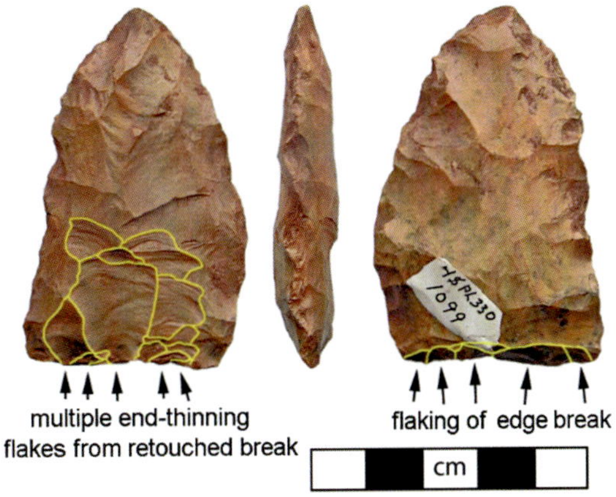

Biface 1099

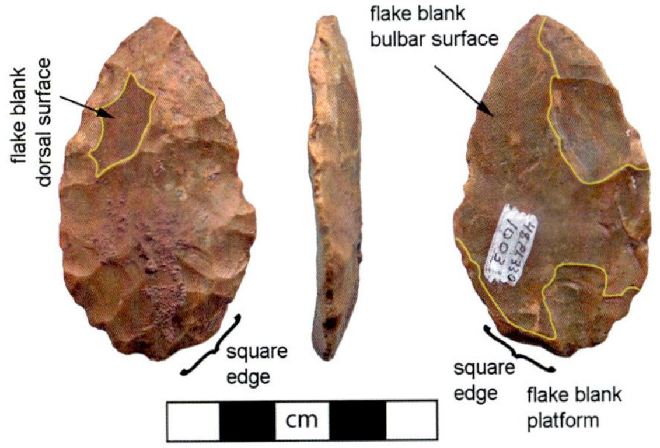

Biface 3388

Biface 1003

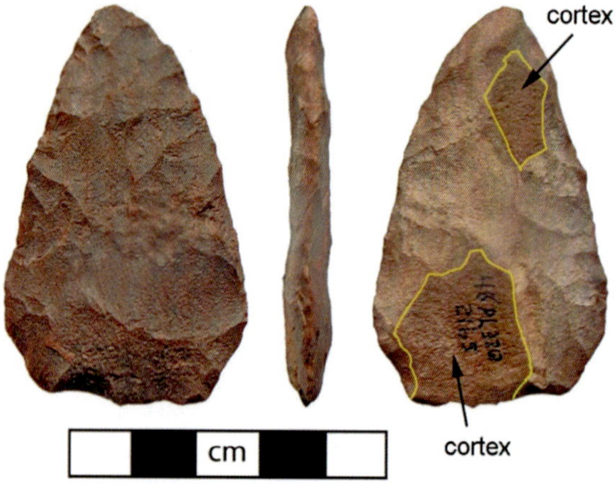

Biface 2165

Biface 1014

Biface 2182

Biface 1207

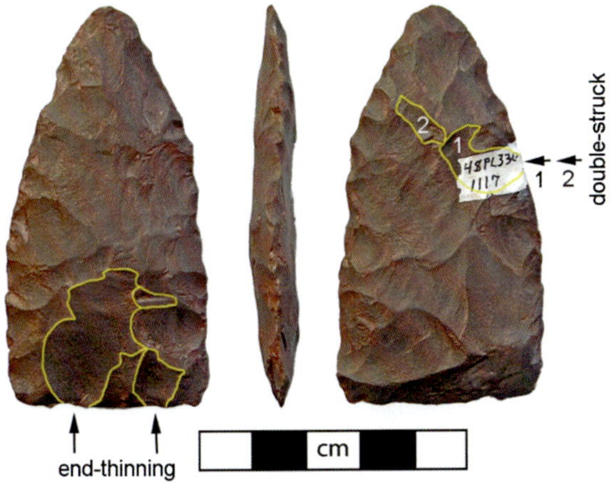

Biface 1117

Biface 645

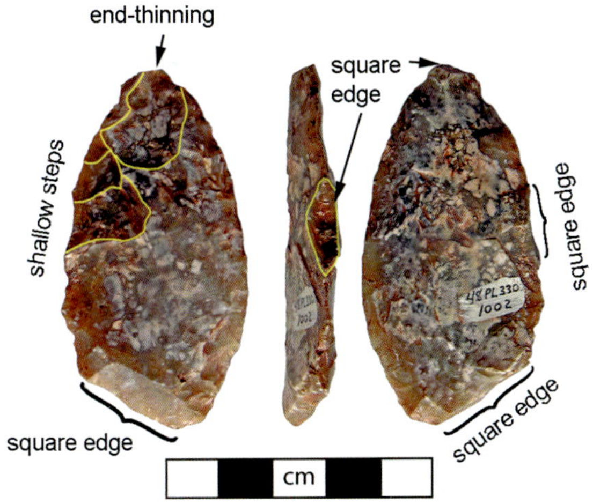

Biface 1002

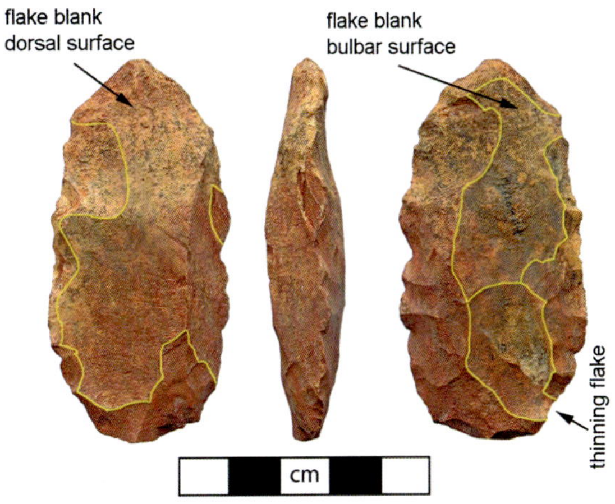

Biface 114

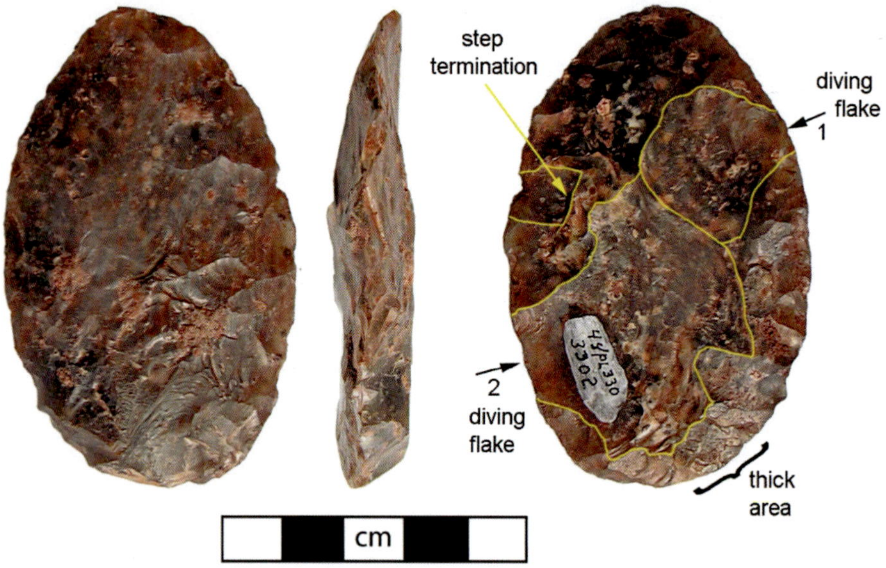
step
termination
diving
flake
1
4SPL330
3302
2
diving
flake
thick
area
cm

Biface 3302

Chapter 7

Appendix 2

Appendix 7- 2. Sunrise Point Analysis

Cat #	Tech	Phase	Material	obverse	reverse	Base obv	Base rev	Base form	Abandonment
41	pth	fin	sst	spt	spt	spt	spt	3	useable
1114	pth	fin	Ark sst	spt	spt	pse flt	spt	25	useable
1115	pth	fin	Krf flt	spt	sps	spt	sps	8	could have been repointed
1116	pth	late	Htv jsp	spt	spt	sps	spt flt	25	could have been repointed
1118	pth	fin	sst	spt	spt	prc	sps	25	useable
1131	pth	fin	por	spt	spt	prs	spt abt	25	useable
1132	pth	fin	opalite?	spt	spt	spt	spt	8	useable
1136	pth	rew	jsp?	spt	spt	prs	prs	38	useable
1138	psh	rew	JV agt	sps	pps	spt	sps	25	useable
1140	pth	fin	por	spt	spt	spt	spt	25	useable
1143	pth	fin	Mnv jsp	spt	spt	prs	prs	25	useable
1144	pth	fin	Htv jsp	spt	spt	prc flt	spt	25	useable
1146	psh	rew	JV cht	pps	spt	sps	sps	7	reuse potential with repaired tip
1147	pth	rew?	JV cht?	spt	sps	spt	spt	3	useable
1968	pth	fin	Mor cht?	spt	spt	prs	spt	25	useable with minor tip work
1969	pth	rew	JV cht	spt	spt	prc lat	sps	25	reuse potential with redone tip
1972	P rc pth	fin	Ark sst	ppt	spt	spt	spt	32	unfinished
1974	pth	fin	Spd sst	spt	spt	missing	missing	25?	exhausted
1976	pth	fin	chl	spt	spt	sps	spt	25	useable
2000	pth	rew	Krf flt	spt	spt	spt	sps	25	reuse potential with repaired tip
2169	pth	fin	Krf flt	spt	spt	spt	spt	25	useable
2170	pth	fin	por	spt	spt	spt	spt	25	useable
2172	pth	fin	Htv jsp	spt	spt	spt	spt	34	reuse potential with redone tip
2173	pth	fin	Spd sst	spt	ppt	?	?	25	exhausted
2176	pth	rew	JV agt	spt	spt	sps	sps	3	useable
2177	pth	fin	Htv jsp	spt	spt	spt	spt	25	reuse potential with redone tip
2179	pth	fin	JV cht	spt	spt	sps	sps	25?	useable
3167	psh	rew	JV jsp	sps	sps	prc flt	sps	25	reuse potential with redone base
3180	pth	rew	Htv jsp	spt	sps	sps	sps	25	useable

Codes:

Technology- prs= pressure; pth= pressure thinning; psh= pressure shaping; prc= percussion; flt= flute; abt= abrupt retouch; lat= lateral; pse= pseudo

Phase- fin= finished; rew= reworked

Material Source- Krf= Knife River; Ark= Arikaree; Htv= Hartville; JV= John Voight; Mnv= Manville; Mor= Morrison; Spd= Spanish Diggings

Material Type- sst= silicified sandstone; opl= opalite; agt= agate; cht= chert; chl= chalcedony; por= porcelinite; flt= flint

Appendix 7-2.2 Sunrise Point Metrics

Cat #	Max L	Max W	Max TH	Base W	Th at Max W	Th at 3/4 base	Max W/Th	Base indent	L at max W	Weight (gr)
41	53	26	6	22.5	6	6	4.3	2.5	17	8.6
1114	43	25.5	4	23	3.8	4	6.4	5	13	5.7
1115	59	28	6	24	6	6	4.7	4.5	23	10.3
1116	63	29	5	26	5	5	5.8	3	23.5	11.4
1118	31	24	5	22	5	5	4.8	3	21	8.8
1131	56	28.5	5.5	22.5	5.5	5.5	5.2	3	24.5	11.6
1132	57	29	5	24	5	4	5.8	3	24	8.2
1136	40	22	5	19	5	5	4.4	1	12	5.5
1138	42	26	6	22	6	6	4.3	2	13	7.3
1140	45	25	5	24	4.5	5	5	5	8	6.9
1143	45	25	5	21.5	5	5	5	2	18	6.6
1144	50.5	29	5	25	5	5	5.8	5	12	8.8
1146	35.5	24	5	22	5.5	5.5	4.8	2	11	5.9
1147	43	22	5	18.5	5	5	4.4	1	20	6.4
1968	47	27	5.5	25	5.5	5.5	4.6	3	15	8.3
1969	55	26	5.5	24	5.5	5.5	4.7	5	17	8.8
1972	49	26	4.5	23.5	4.5	4.5	5.2	3	14	7.7
1974	54	n/a	5	n/a	n/a	5	n/a	n/a	n/a	7.6
1976	48	27	5	24.5	5	5	4.9	1.5	16	6.7
2000	61	26	6	22	6	6	4.3	3	18	11.6
2169	46	27	4.5	21	4.5	4.5	6	5	22	7.6
2170	57	26	5	24	5	5	5.2	2.5	22	7.9
2172	43	24	5.5	22	5	5	4.4	3	15	6.2
2173	53	28	5	26	5	5	5.6	7	16	7.5
2176	43	24	5	21	5	5	4.8	1	20	6.7
2177	53	26	5	23	5	5	5.2	3	13	7.9
2179	61	30	5	20	5	5	6	2	28	12.3
3167	62	29	5.5	25	5.5	5	5.3	5	15	12.3
3180	46	26.5	6	24	5.5	6	4.4	2.5	22	8.6

Dimensions in millimeters

Appendix 7-2.3 Powars II (48PL330) Point Notes

These notes are intended to include observations beyond what was included in the spreadsheet, eg. color, whether or not basal flaking was truncated by lateral flaking, areas of margin grinding, etc.

Object#	Color; Condition (including patina, ochre, weathering); Technology (blank, phases, retouch, margin grinding, reworking); Use; Other
41	pink; small bit of tip and one ear missing by recent damage, some chemical erosion, ochre coating; blank and early phases unknown, wide spacing on pressure thinning flakes, many step terminations, Basal flaking after lateral; stack at end of basal flake scar on obverse, this is the point of maximum width, thickness and thickness at ¾ base, otherwise a thin point, thickness just distal to stack is 4.5with width at 24 (w/th= 5.3), no retouch or reworking, margin grinding not evident but may have been eroded off; no evidence of use; would have been a functional point.
45	Yellow-tan with light brown tip; complete but tip reworked, ochre stained; blank and early phases unknown, pressure flake scars do not pass the midline, basal flaking invaded by lateral flake scars, minimal selective pressure retouch, pressure flaking, light margin and base grinding to widest dimension. No evidence of reuse.
1114	Dark brown; complete, ochre stained; possible flake blank, pressure thinning, wide spacing, pseudo flute on obverse possible thermal spall negative, selective, noncontinuous pressure retouch on reverse, all over wind polishing? margin grinding not evident (possibly because of surface polishing); no evidence of use or reworking, retains distal needle point.
1115	Beer bottle brown, translucent; distal 1/3 missing by major impact flute on obverse, ochre stained; blank and early phases unknown, selective pressure thinning with wide, deep scars on obverse and narrower on reverse, last flaking on both faces on left margin, what looks like retouch is probably margin-long platform preparation with short abrupt beveling, base thinning followed by marginal flaking both faces, light margin and base grinding. Heavy tip impact evident on both faces; no reworking.
1116	brown with yellow tip, dendritic; tip missing, snap break on internal flaw, manufacture break, reverse lightly patinated, ochre stained; late phase, wide, selective pressure thinning, flute on reverse looks like heavy pressure thinning flake scar, could be percussion, may have been final flake removal and end shock removing tip on flaw, no retouch, lower right dorsal margin beveled with abrupt flaking- looks like margin preparation for further flaking, no margin grinding; unfinished, could easily have been retipped. Very nicely made and reason for abandonment unknown.
1118	Dark red; complete except there is asymmetry near the tip, obverse lightly wind polished, ochre stained; selective pressure thinning near base on reverse face, otherwise mainly pressure-shaped, some selective pressure retouch removing some deltas, basal flaking mostly removed by later flaking both faces, obverse has percussion flake scar remnant, left lateral obverse margin lightly ground, otherwise no edge grinding.; possible lateral tip impact, reflaked by pressure shaping.
1132	White and red opalite? Right obverse ear missing, heavily patinated(?), ochre stained, possibly lightly wind polished both faces both faces, wide selective pressure flaking, discontinuous non-invasive, selective retouch (delta removal), obverse basal scar removed by later flaking, reverse basal scar after lateral flaking, indented base offset to remaining ear as if done after break, no apparent edge grinding; no evident use but useable as is.
1137	Maroon; complete, wind polish both faces, selective pressure flaking not reaching the midline, possibly percussion scar remnants at midline but weathering obscures them on both faces, obverse base flake scar removed by later flaking, reverse base either percussion or selective pressure thinning, no evident retouch, lateral margins and base lightly ground; useable but 'blunt'.

1140	Maroon; complete, finished, chemically weathered, ochre stained; obverse medium wide selective pressure shaping, reverse wide pressure thinning, no retouch, both bases pressure thinned, and scars partly removed by further lateral flaking, margin grinding not apparent, but weathering may have obscured it. Useable.
1143	dark red distal ¾ yellow-tan proximal ¼, dendritic; complete with needle point, fresh appearance, basal pressure flaking truncated on both faces by lateral flaking, no retouch, very light grinding on lateral margins and base. Useable.
1144	Yellow ochre color; complete, some lateral damage left obverse margin, fresh appearance, very slight ochre staining, obverse percussion flute, all basal flaking done after lateral flaking, percussion scar remnants in center of reverse face, wide flake spacing except for tip area where it is narrow and fine; Edge damage probably post-depositional. Useable.
1968	Mottled gray basal 2/3 yellow ochre color tip 1/3; very small portion of tip missing, probably loss of needle point; medium to wide selective pressure thinning both faces, basal flaking after lateral flaking, reverse one long pressure thinning scar and several smaller ones, no retouch, medium grinding lateral margins, and base; useable with minimal tip work.
1969	Mottled yellow ochre; tip missing for unknown reason, right obverse margin burin scar, reworked after burin, light ochre staining; percussion scar remnants both faces, medium width pressure flake scars, no retouch, medium grinding lateral margins, and base; burin scar probably from impact, tip and lower ear flaked after burin removal, could be reused with minimal tip flaking.
1972	Dark red; light wind polishing? percussion scar ending in step fracture on obverse followed by selective pressure thinning, right obverse margin beveled for platform preparation for reverse pressure flaking, no retouch or margin grinding; could be finished and used. Asymmetrical base and tip not finished.
1976	white translucent; complete except for small nick obverse right margin, probably post-depositional, reverse has thin white patina; wide to narrow pressure thinning both faces, one serial sequence of 4 scars obverse distal 1/3, tip to base, no retouch but short bevel obverse left margin, platform preparation- not retouch, obverse base pressure shaping after lateral flaking, reverse pressure thinning after lateral flaking, no margin or base grinding, retains needle tip; Useable.
2169	beer bottle brown; moderate white patina on reverse; obverse percussion scar distal ½, medium width pressure scars both faces, obverse base pressure thinning, reverse base pressure thinning with one long central scar, base flaking after lateral flaking both faces, continuous fine retouch obverse left margin, but short platform prep bevel right margin, needle tip, light marginal grinding but not on base. Useable.
2170	Maroon; chemical weathering all over including dulled margins; selective pressure thinning both faces and both bases, obverse base thinning truncated by lateral flaking, reverse base thinning after lateral thinning, reverse left margin has slight abrupt bevel for platform preparation- not retouch; margin and base grinding, but also weathered; useable.
2172	Reddish-brown; reverse lightly patinated, tip missing- probably just a needle point, bend, unknown cause; Obverse pressure shaping but multiple pressure thinning base flake scars, originating after lateral flaking, reverse pressure thinning including base but lateral flaking after base thinning, no retouch, no margin or base grinding; useable after minor tip repair.
2173	Brown-tan; tip and obverse left ear missing, snap breaks of unknown origin, light ochre staining; obverse wide pressure thinning, base probably pressure thinning, reverse remnant of diving percussion flake scar in middle, pressure thinning at distal end, some beveling for platform preparation, not retouched; no potential for reuse as a point.

2177	Yellow ochre; tip crushing and small impact flute on obverse, obverse wide selective pressure thinning, reverse medium width pressure thinning, same for base both faces, obverse lateral flaking after base thinning, reverse base thinning after lateral flaking, some possible minimal non-invasive, discontinuous retouch, heavy grinding on margins but very light or missing on base; enough remains that the tip could be reflaked.
2179	Brown; intact including needle tip, light ochre staining; both faces wide widely spaced pressure thinning, some beveled edge platform preparation, continuous non-invasive bifacial retouch from widest dimension to tip, light margin and base grinding. Useable.
3167	Brown and ochre yellow; obverse left ear missing by bend break, significant ochre residue; obverse selective pressure shaping but flat with medium to narrow spacing, reverse pressure shaping with medium spacing, obverse base has percussion flute scar terminating in a step after bifacial, continuous, non-invasive retouch on right side and in indented base, reverse base pressure shaping before and after lateral flaking and obverse retouching, left lateral edge from tip down 1/3 of length bifacially pressure shaped as a reworking event, heavy grinding on obverse right margin but not left or base. Could have been rebased for reuse. Basal fluting and all retouching probably part of reworking of damaged point.
3180	Reddish brown with dendrites; no patina, ochre or weathering; obverse selective pressure thinning with narrow spacing, removal of step termination on internal flaw by flake from opposite margin, remnant of flake scar which was probably an impact, tip retouched bifacially, reverse pressure thinning, narrow scars, obverse base one pressure thinning and several pressure shaping removals after lateral flaking, reverse pressure thinning after lateral flaking, margins and base lightly ground. Useable.
3910-908	light gray translucent; light ochre stain; obverse selective pressure shaping, narrow scars, remnant of long pressure thinning flute after lateral flaking but then selective pressure shaping of base, reverse one percussion scar ending in deep step, right margin, percussion remnant adjacent from left margin, then selective pressure shaping with narrow scars, some discontinuous, selective, non-invasive retouch, especially in hafting area, very light margin and base grinding; useable.
3923-724	Light tan-yellow; obverse left ear missing by snap break, left margin has crushed area, reddening around margins may be thermal oxidation after ear break; obverse irregular mix of pressure shaping and thinning, base has pressure thinning scar before lateral flaking, continuous, non-invasive selective retouch on right margin, crushed area from obverse on left margin, reverse selective pressure thinning diagonal upper left to lower right from left margin some scars full-face, base selective pressure thinning, continuous non-invasive bevel retouch distal 1/3 right margin, continuous non-invasive retouch proximal 2/3 right margin, right margin heavily ground 61% distance base to tip, left margin damaged. Could be rebased.
3936-641 B	Light brown; tip missing with two sharp bend breaks, post-depositional?, light patina obverse, light ochre staining reverse; obverse selective pressure thinning, well-spaced medium width scars, base selective pressure shaping truncated by lateral flaking, reverse selective pressure thinning, well-spaced wide to medium width scars, base selective pressure shaping truncated by lateral flaking, no retouch, light margin and base grinding; use potential with retipping.
1124/1125	reddish-brown translucent; broken near the middle, snap break probably post-depositional, ochre staining; obverse selective pressure thinning, well-spaced wide scars, some serial sequences of 2-3 flake removals, base flaking removed by lateral flaking, no retouch, dominant deltas, reverse selective pressure thinning, well-spaced medium wide scars, some serial sequences of 2-3 flake removals, base flaking selective pressure thinning, partly removed by lateral flaking, no retouch, no margin or base grinding. Unbroken would have been easily useable and finished with minimal effort.

1131 Maroon; obverse left ear small portion missing, two abrupt flake scars, three small edge dings, one fresh, unknown reason, all-over chemical weathering; both faces selective pressure thinning, not retouched but some areas of minimal beveling for platform preparation, obverse base pressure but scars truncated by lateral flaking, reverse base long pressure thinning scar then some lateral flaking and abrupt pressure retouch, heavily ground margins but unground base; useable.

1974 Tan-orange; obverse right ear and part of midsection missing, break from just inside left ear, perverse and unifacial flake scars, unknown cause, reverse lightly patinated, some ochre residue; obverse selective pressure thinning, medium to wide scars, reverse selective pressure thinning medium to narrow scars, no retouch either face, light margin grinding; exhausted no further point use potential.

823 Cream. Obverse left ear missing by horizontal burin removal, tip by steep flake, light ochre staining; Obverse percussion followed by steep selective pressure shaping, reverse the same but less steep, discontinuous selective pressure retouch both faces and margins, non-invasive, very light margin grinding but not base; could easily be retipped.

3935-74 Yellow ochre with dendrites; complete and finished; percussion preform with diving flake removals both faces, tip produced with selective pressure shaping both faces, obverse formed with selective no-invasive pressure and non-invasive retouch, reverse formed by continuous selective, non-invasive retouch, obverse base selective pressure shaping, reverse base remnant of percussion flake scars and non-invasive pressure retouch, needle tip, margins lightly ground but base unground; useable. Deep step termination on obverse left margin leaving a small square edge.

2000 Beer bottle brown; minor tip break, flake scar originating from reverse, remnant of long burin removal from obverse left ear, distal margin and ear reworked after spall removal, unknown break reason, possible remnant of tip impact on obverse but reworked, probably use damage, light ochre staining on reverse; both faces selective pressure thinning with wide scars, obverse base selective pressure thinning after lateral flaking, reverse base selective pressure shaping truncated by lateral flaking, minimal selective discontinuous non-invasive retouch both faces, distal obverse margin retouch with steep widely space pressure flaking to repair burin damage , light grinding on margins and base; useable with minimal tip reworking.

1119 Brown; medium quality stone; minimal tip missing (needle) snap, obverse left ear minimal missing flake, light ochre staining; obverse selective pressure thinning one short deep step termination, reverse percussion scar then selective pressure shaping with small stack exactly opposite deep step on obverse, no retouch but small bevels on lower margins to set up platforms, obverse base either percussion or large pressure thinning scar truncated by lateral flaking, reverse base pressure thinning and abrupt pressure retouch, heavy margin grinding but sharp base. Useable with minimal tip sharpening.

1926-106 Mottled pink and white; some minor damage/crushing obverse left distal, possible recent impact damage (looks fresh), possible light patina both faces; selective pressure shaping , medium width running mostly to midline, on pressure thinning removal from right margin 1/5th distance from base, selective pressure retouch, invasive and mainly directly under previous flake scars giving the right margin an irregular serration, more even continuous retouch on left proximal margin, reverse more irregular selective pressure shaping with medium to narrow scars, continuous short invasive pressure retouch on right margin, obverse base two medium wide pressure shaping scars after lateral flaking, reverse multiple pressure shaping scars truncated by lateral flaking, obverse left margin and base lightly ground, right margin not ground; useable.

1138 Clear with red inclusions; ochre stained; obverse selective pressure shaping, several diagonal scars upper left to lower right, continuous non-invasive pressure retouch proximal ½, reverse percussion scars followed by selective pressure shaping, retouch along proximal margins, obverse base end of one pressure thinning flake scar then non-invasive pressure retouch, reverse base selective pressure shaping, no grinding of margins or base; useable. Looks to have been rebased with pressure retouch on margins and pressure shaping on base.

1146 Reddish-brown; tip missing with flake removal, ochre staining; obverse percussion flake scar followed by selective pressure shaping, continuous abrupt pressure retouch along left margin, reverse selective pressure shaping, continuous abrupt pressure retouch along right margin, obverse base pressure shaping, reverse base abrupt pressure shaping, margins and base not ground. Potential to be used if tip resharpened. Basar ears have remnants of previous lateral break. Rebased point.

2176 White translucent; heavily patinated all over; obverse selective pressure thinning except distal are is selective pressure shaping, fine discontinuous non-invasive retouch proximal margins (stem), reverse selective pressure thinning, continuous non-invasive pressure retouch on distal margins, obverse base pressure thinning truncated by lateral flaking, reverse base pressure shaping, margins lightly ground base sharp; useable. Retipped Goshen point.

1126 Creamy yellow with dendrites; small portion of tip missing with flake scar, unknown reason, light ochre staining; obverse remnant of ventral surface of flake blank, bulb of percussion at base, bulb removed by percussion thinning flake, distal end selective pressure shaping, no retouch but remnants of margin platform preparation, reverse selective pressure shaping some diagonal from upper left to lower right, no retouch, obverse base pressure shaping, reverse base pressure thinning, no margin or base grinding; Useable with minimal tip sharpening.

1965 Beer bottle brown; white patina on reverse; both faces selective pressure shaping, very narrow scars, discontinuous invasive pressure retouch, obverse base single pressure thinning scar, reverse two pressure thinning scars, light grinding on margins, sharp base; distal of 'stem' possibly reworked; useable. Retipped Agate Basin base?

1123 Dark yellow-brown with dendrites; obverse left lateral damage at ¾ length, impact?; selective pressure shaping, continuous selective pressure retouch some abrupt some invasive, reverse percussion flake scar distal portion then selective pressure shaping, abrupt continuous pressure retouch both margins from base to widest point, selective pressure retouch on distal margins; Obverse base percussion flute with end removed by lateral pressure shaping and sides modified by abrupt retouch, reverse base two basal thinning flakes one could be percussion the other is pressure thinning with ends removed by lateral pressure shaping and sides modified by abrupt retouch, medium margin grinding but sharp base; useable. Unidentified fluted point, with possible tip reworking and use damage.

1964 Yellow ochre with dendrites; very tip minimally missing probably lost needle point, unknown reason, obverse heavily patinated; obverse selective pressure shaping with minimal non-invasive selective pressure retouch, reverse has percussion scar followed by selective pressure shaping and a small amount of selective pressure non-invasive retouch, obverse base percussion thinning scar truncated by lateral flaking, reverse base short percussion shaping scar, light grinding on margins but sharp base; useable. Plainview.

1147 Dark reddish-brown; burned, reverse lightly patinated, ochre staining/residue; obverse selective pressure thinning, some possible discontinuous non-invasive retouch but could also be edge damage, reverse selective pressure shaping, obverse base one relatively wide pressure thinning scar, reverse multiple pressure thinning scars truncated by lateral flaking, margins lightly ground base sharp; useable. Asymmetrical form may indicate reworking. Sunrise?

1135 Tan; tip missing from impact flake removal, obverse right ear burinated from lateral margin to center of base, both probably impact breaks, ochre stained; both faces with comedial selective pressure shaping and discontinuous selective pressure retouch mainly removing deltas, medium grinding on margins and base; exhausted. Unfluted Folsom?

1134 Tan; tip missing a small piece impact flake on obverse, obverse left ear missing small piece by snap, light ochre staining, 2 pieces refit- recent break?; obverse widely spaced selective pressure thinning flakes, margins have continuous non-invasive pressure retouch, reverse widely spaced parallel-sided percussion flakes with abrupt selective discontinuous pressure retouch mainly removing deltas, some damage on reverse right margin, obverse base one shallow pressure thinning flake then non-invasive pressure retouch, reverse base non-invasive pressure retouch truncated by lateral flaking, medium grinding on margins and base. Coul have been retipped and used (assuming main break was post-depositional.

1971 Dark red; small amount of tip missing by flaking- needle?, ochre staining; obverse mostly percussion scars with selective pressure thinning, discontinuous selective non-invasive retouch both margins, reverse a couple percussion scar remnants but mainly selective pressure shaping, a section of small invasive pressure retouch in central area of right margin, obverse base pressure shaping, reverse base pressure thinning, margins medium ground, base lightly ground; could be useable with minimal tip resharpening. Plainview.

1970 Maroon; minimal chemical weathering/wind erosion, light ochre staining; both faces selective pressure shaping with selective invasive pressure retouch mostly removing deltas, obverse base pressure thinning after lateral flaking, reverse base pressure shaping after lateral flaking, medium margin grinding light base grinding; useable. Plainview.

3404 Yellow ochre; complete but unfinished; flake blank with remnant of platform at tip, obverse selective pressure shaping with scars ending at a midline offset to the left, no retouch, reverse mostly flake blank ventral surface, selective pressure thinning at base (to eliminate curve) and tip (to reduce bulb), no retouch, obverse base pressure shaping, reverse base pressure but mostly eliminated by lateral flaking, no margin or base grinding. Unfinished but with potential to be. Plainview?

1158 Reddish-pink; complete with light patina on both faces; flake blank with platform at base, obverse mostly flake blank dorsal surface with scars parallel to flake axis and originating at the platform end, probably a biface thinning flake, right side about ½ original 'square edge' flake scar, moderately patinated, minimal selective pressure shaping along left margin with no flakes reaching the midline, selective pressure shaping on right margin from base and tip about ¼ length each, needle tip, obverse base pressure shaping, reverse base pressure shaping, no margin or base grinding; useable. Plainview?

Appendix 2.4 Sunrise Point Images

48-PL-330-41

48-PL-330-1114

48-PL-330-1115

48-PL-330-1116

48-PL-330-1118

48-PL-330-1131

48-PL-330-1132

48-PL-330-1136

48-PL-330-1138

48-PL-330-1140

48-PL-330-1143

48-PL-330-1144

48-PL-330-1146

48-PL-330-1147

48-PL-330-1969

48-PL-330-1972

48-PL-330-1976

48-PL-330-2000

48-PL-330-2169

48-PL-330-2170

48-PL-330-2172

48-PL-330-2176

48-PL-330-2179

48-PL-330-3167

48-PL-330-2177

48-PL-330-2173

48-PL-330-1974

Chapter 7

Appendix 3

Powars II Sunrise Points (n=34)

Main technology (n=33)
pressure thinning= 28
pressure shaping= 3
percussion with pressure thinning n=1
percussion with pressure shaping n=1

Technology facial combinations (n=32)
pressure thinning/pressure thinning n= 21
pressure thinning/pressure shaping n=3
pressure shaping/pressure shaping n=1
other n=7

Technology of bases (n=67)	**obverse**	**reverse**
pressure thinning	15	16
pressure shaping	7	12
pressure (not further specified)	4	2
fluting percussion 3	0	
fluting pressure		1
percussion	1	
unknown	2	2

	N	Range	Mean	Median	Mode
Length	24	31-62	48.7	48	43/35
Width	30	22-31	27.1	26	26
Thickness	33	4-7.5	5.3	5.0	5.0
W/Th	32	4.1-6.4	5.0	5.0	5.2
Base W	27	18.5-26	22.4	22	22/24
Base indentation	31	1-5	2.9	3	3
Weight (Grams)	22	5.5-15.5	8.2	7.6	7.6/8.6

Measurements (mm)

Condition
Complete n=18
Incomplete n=14 Portions: distal= 2 Proximal= 12
Fragment n=1

Abandonment assessment
useable n=20 (9 possibly reworked)
could be finished or rejuvenated n=11
exhausted n=2

Definitions

Main technology is the dominant process that was used and is visible on the finished piece. This should not be confused with primary technology, which indicates the initial flaking process(es). When more than one dominant process is present, for example pressure thinning and pressure shaping, the most difficult to achieve takes precedence, in this case pressure thinning. Another example is percussion thinning and percussion shaping. The former is more specialized and difficult to accomplish so a piece that exhibits both would be considered to have a percussion thinning main technology.

Pressure thinning is when flakes travel past the midline on a flat trajectory. Another way of saying this is flat transmedial. Its primary result is either retaining or producing a flat face. In W/Th ratios this may not actually result in relative thinning, but also does not produce relative thickening. This method often results in shallow step terminations with 'hangnails' common. *Hangnails* are the ends of flakes that terminated in step fractures leaving a piece still attached to the flaked surface.

Pressure shaping is either non-invasive or invasive but tends not to pass the midline and if it does it has a curved trajectory. Generally, it either retains or produces a convex surface. It also tends to result in relative thickening. Flake scars usually have feather terminations.

Retouch is defined as the process of **final** margin adjustment which results in regularization of the margin but may or may not affect the face of the piece. Retouch may be pressure or percussion, continuous or selective, abrupt or invasive. It may be unifacial or bifacial and it may be applied along entire margins or just in certain sections/areas. In some cases, what looks like retouch is the remnant of platform preparation, indicated by subsequent flaking.

Deltas are the areas along the margin between flake scars. These commonly occur where adjacent flake scars are spaced so that a triangular area of previous flaking is isolated, resulting in a 'wavy' edge.

For other technological definitions see Bradley et al 2010:56-106.

Incomplete is where piece(s) are missing, but the original outline and dimensions may be reasonably estimated.

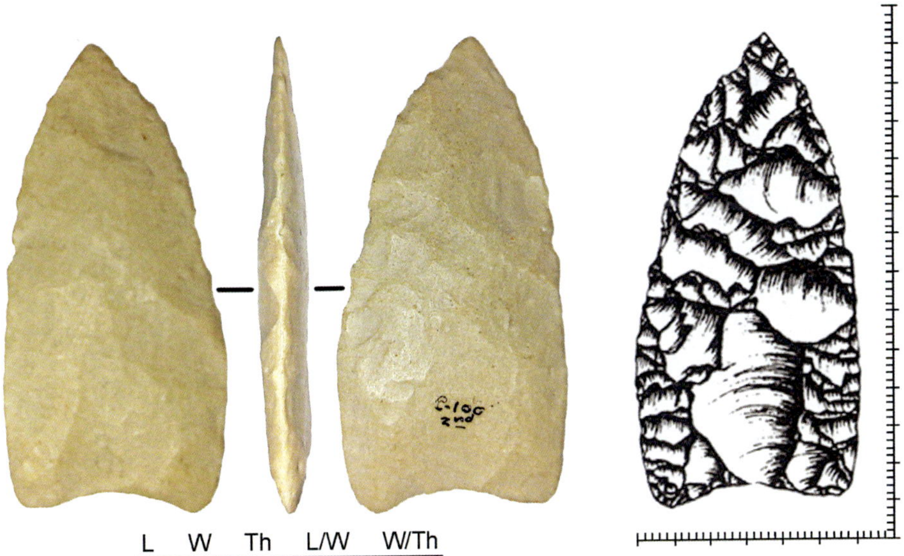

L	W	Th	L/W	W/Th
59	29.5	7	2/1	4.2/1

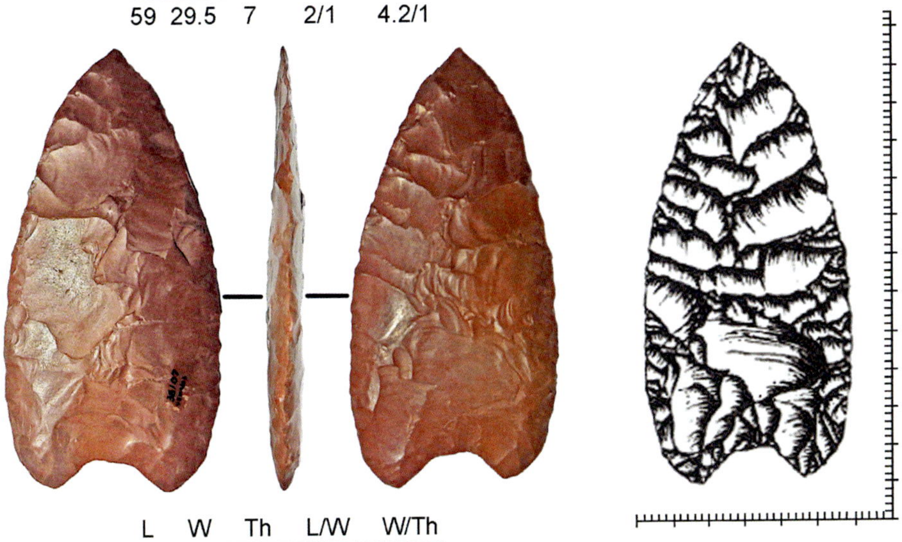

L	W	Th	L/W	W/Th
56	28	4.5	2/1	6.2/1

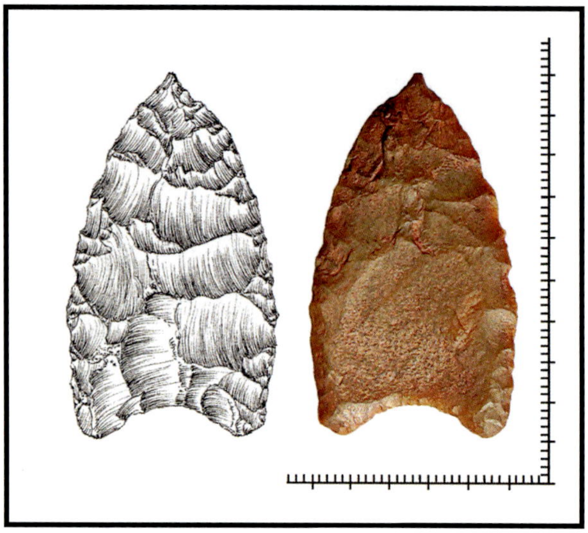

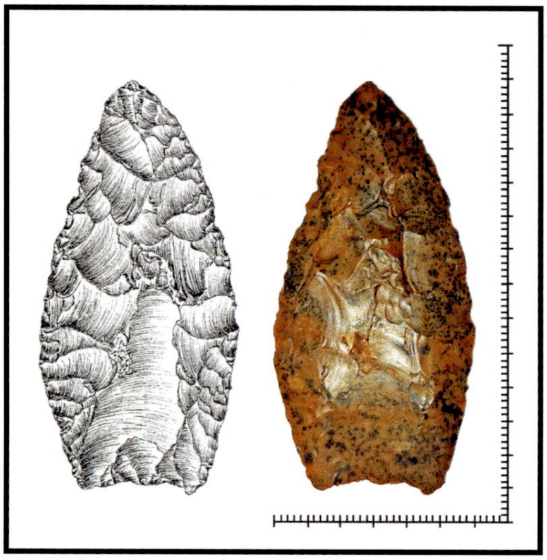

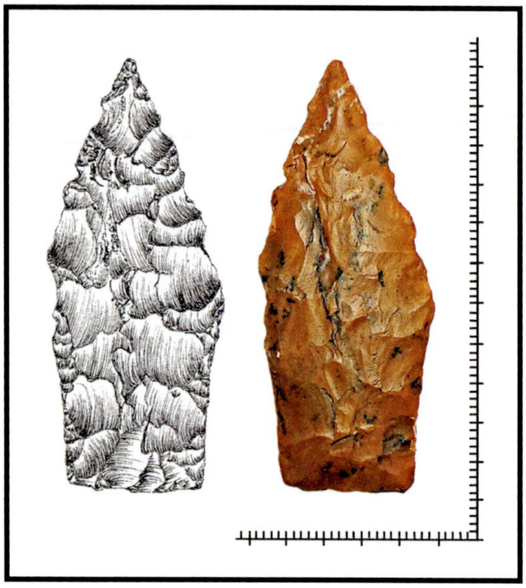

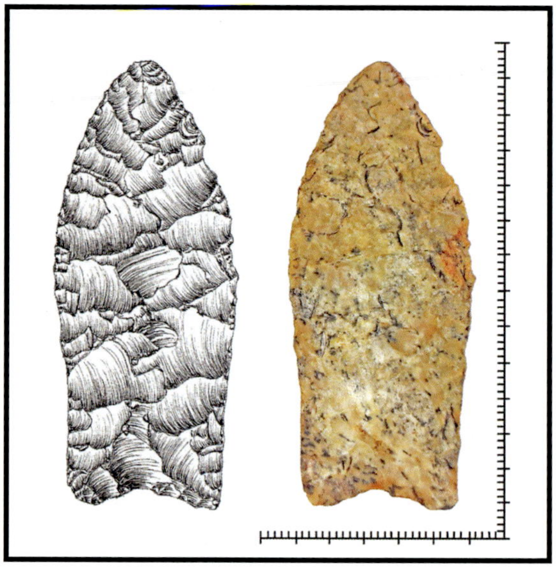

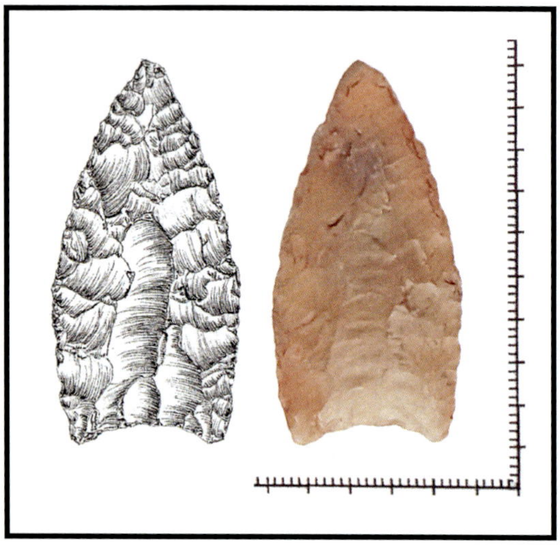

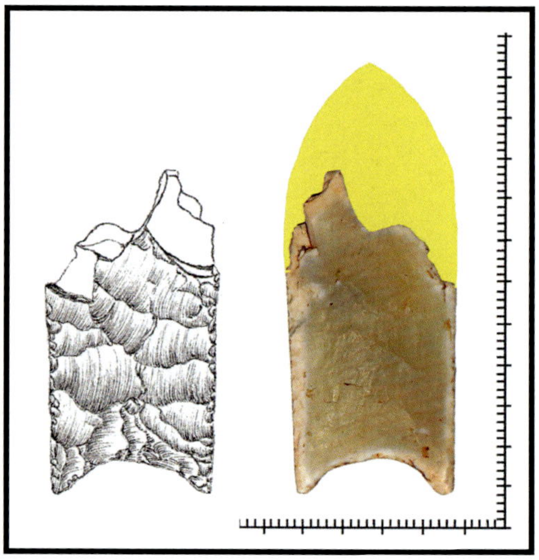

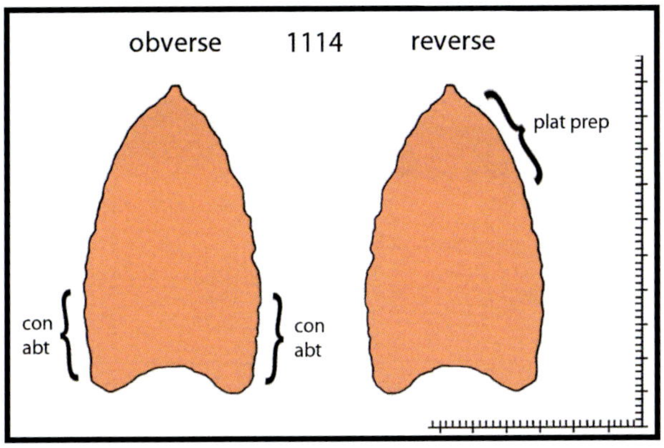
obverse
1114
reverse
plat prep
con
abt
con
abt

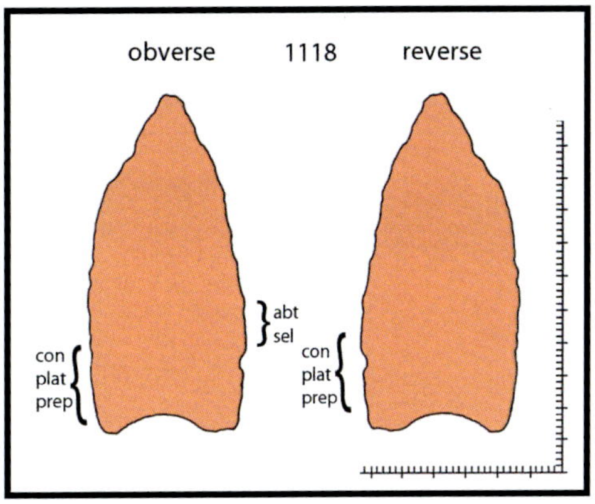
obverse
1118
reverse
abt
sel
con
plat
prep
con
plat
prep

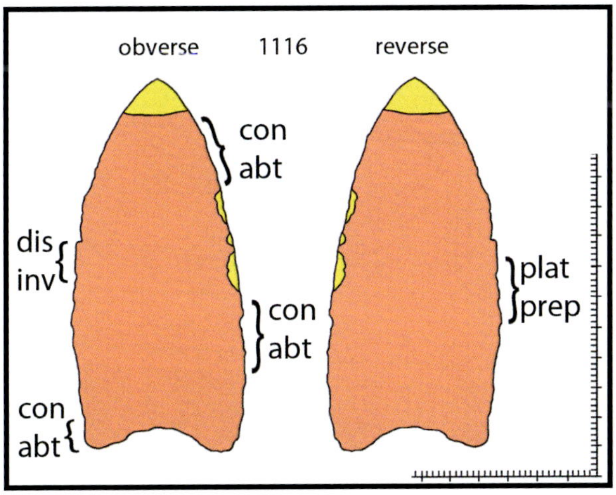
obverse
1116
reverse
con
abt
dis
inv
con
abt
con
abt
plat
prep

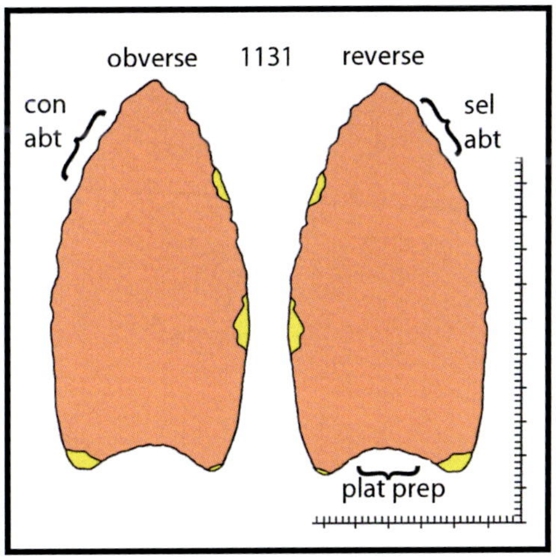
obverse
1131
reverse
con
abt
sel
abt
plat prep

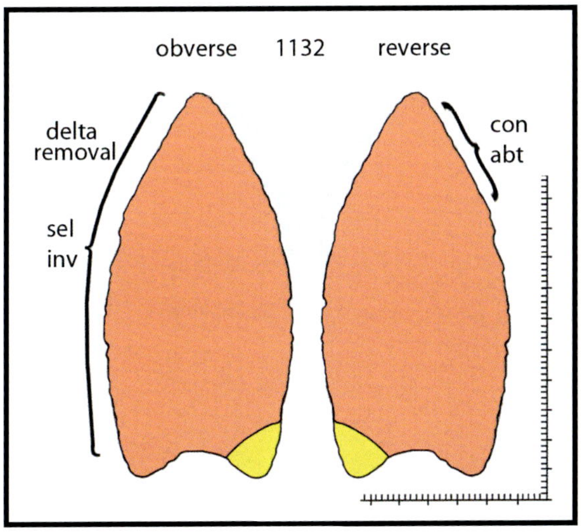
obverse
1132
reverse
delta
removal
sel
inv
con
abt

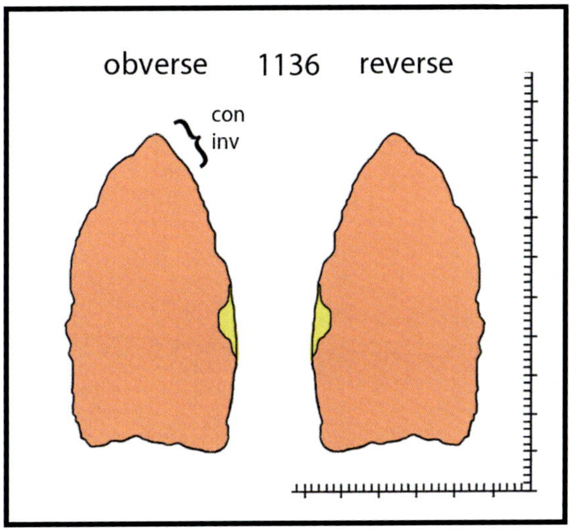
obverse
1136
reverse
con
inv

Projectile Points

Cat #	Main technology	Condition	Portion	Break	Cause	Phase	Material	obverse	reverse
41	pressure thinning	incomp	90%	tip and corner	fresh	finished	silicified sandstone	spt	spt
45	reworked	comp	n/a	n/a	n/a	reworked	Smokey Hills jasper?	sps	sps
823	percussion, pressure shaping	incomp	1 ear and tip missing small pieces	both flake scar	?	finished reworked?	chert	pps	pps
1114	pressure thinning	comp	n/a	n/a	n/a	finished	Arikaree silicified sanstone	spt	spt
1115	pressure thinning	incomp	base and midsection	impact flute	heavy impact	finished	KRF	spt	sps
1116	pressure thinning	incomp	base midsection	tip and corner	internal flaw	late	Hartville jasper	spt	spt
1118	pressure thinning	comp	n/a	n/a	n/a	finished	silicified sandstone	spt	spt
1119	pressure shaping	comp	n/a	n/a	n/a	finished	Spanish diggings ss	ppt	pps
1123	percussion, pressure shaping	comp	n/a	n/a	n/a	finished	Hartville jasper	sps	pps
1126	percussion, pressure shaping	incomp	small portion of tip missing	flake scar	?	finished	opalite?	pps	sps
1131	pressure thinning	incomp	one ear partly missing	flaked	?	finished	porcelinite	spt	spt
1132	pressure thinning	incomp	one ear missing	bend	?	finished	opalite?	spt	spt
1134	percussion, pressure thinning	incomp	tip missing, ear missing	impact flake scar	use as projectile	finished	Spanish diggings ss	spt	ppt
1135	pressure shaping	incomp	tip missing, ear burinated	impact flake scar	use as projectile	finished	JV chert?	sps	sps
1136	pressure thinning	comp	n/a	n/a	n/a	finished-reworked	jasper?	spt	spt
1137	pressure shaping	comp	n/a	n/a	n/a	finished	porcelinite	sps	sps
1138	pressure shaping	comp	n/a	n/a	n/a	finished reworked	JV agate	sps	pps
1140	pressure thinning	comp	n/a	n/a	n/a	finished	porcelinite	spt	spt
1143	pressure thinning	comp	n/a	n/a	n/a	finished	Hartville jasper Manville variety	spt	spt
1144	pressure thinning	comp	n/a	n/a	n/a	finished	Hartville jasper	spt	spt
1146	pressure shaping	incomp	small portion of tip missing	flake scar	?	finished reworked	JV chert	pps	spt
1147	pressure thinning	comp	n/a	n/a	n/a	finished reworked?	JV chert?	spt	sps
1158	pressure shaping	comp	n/a	n/a	n/a	finsihed	JV chert	sps	n/a
1964	percussion, pressure shaping	comp	n/a	n/a	n/a	finished	Hartville jasper	sps	pps
1965	pressure shaping	comp	n/a	n/a	n/a	finished reworked	Knife River flint	sps	sps
1968	pressure thinning	incomp	minimal tip missing	snap	?	finished	Morrison chert?	spt	spt
1969	pressure thinning	incomp	tip missing	bend	?	reworked	JV chert	spt	spt
1970	pressure shaping	comp	n/a	n/a	n/a	finished	porcelinite	sps	sps
1971	percussion, pressure shaping	incomp	tip missing	flaking	?	finished	Arikaree silicified sanstone	pps	pps
1972	perc and press thinning	comp	n/a	n/a	n/a	late?	Arikree silicified sandstone	ppt	spt
1974	pressure thinning pressure thinning	frag	one ear and edge of midsection missing	perverse and flaking	?	finished	Spanish Diggings ss	spt	spt
1976	pressure thinning	comp	n/a	n/a	n/a	finished	chalcedony	spt	spt
2000	pressure thinning	incomp	small portion of tip missing	flake scar	?	finished reworked	Knife River flint	spt	spt
2169	pressure thinning	comp	n/a	n/a	n/a	finished	Knife River flint	spt	spt
2170	pressure thinning	comp	n/a	n/a	n/a	finished	porcelinite	spt	spt
2172	pressure thinning	incomp	tip missing	snap	?	finished	Hartville jasper	spt	spt
2173	pressure thinning	incomp	tip and one ear missing	snap breaks	?	finished	Spanish diggings ss	spt	ppt
2176	pressure thinning	comp	n/a	n/a	n/a	finished reworked	JV agate	spt	spt
2177	pressure thinning	incomp	tip missing	impact flute	impact	finished	Hartville jasper	spt	spt
2178	percussion and pressure shaping	incomp	corner missing	snap	?	finished	porcelinite	pps	pps
2179	pressure thinning	comp	n/a	n/a	n/a	finished	JV chert	spt	spt
3167	pressure shaping	incomp	obv left ear missing	bend	?	reworked	JV jasper	sps	sps
3180	pressure thinning	comp	n/a	n/a	n/a	reworked	Hartville jasper	spt	sps
3404	pressure shaping	comp	n/a	n/a	n/a	Phase 2	Hartville jasper	sps	spt
1124/1125	pressure thinning	comp	2 refit pieces	snap	post depositional	late phase	JV agate	spt	spt
1926-1064	pressure shaping	comp	n/a	n/a	n/a	finished	Alibates dolomite	sps	sps
3910-908	pressure shaping	comp	n/a	n/a	n/a	finished	JV agate	sps	sps
3923-724	pressure thinning	incomp	ear missing and edge damage	snap, crushing	use?	finished	Smokey Hills jasper?	sps	spt
3935-74	percussion, pressure shaping	comp	n/a	n/a	n/a	finished	Hartville jasper	ppr	ppr
3936-641 B	pressure thinning	incomp	tip missing	bend-double	post depositional?	finished	JV chert?	spt	spt

Base obv	Base rev	Base form	Abandonment	Max L	Max W	Max TH	base w	Th at Max W	Th at 3/4 base	Max W/Th	base indent	l at max w	Weight	Affiliation
spt	spt	3	useable	53*	26	6	22.5*	6	6	4.3	2.5*	17	8.6	Sunrise
sps	spt	7	useable	45	26	7	21	7	6	4.3	2	23	8	Plainview
perc flute	sps	7	reuse potential with redone tip	60*	29	7.5	24	7	7	3.9	1*	26	16.4+	Clovis?
pseudo flute	spt	25	useable	43	25.5	4	23	3.8	4	6.4	5	13	5.7	Sunrise
spt	sps	8	could have been repointed	59*	28	6	24	6	6	4.7	4.5	23	10.3+	Sunrise
sps	spt-flute	25	could have been repointed	63*	29	5	26	5	5	5.8	3	23.5	11.4+	Sunrise
percussion	sps	25	useable	31	24	5	22	5	5	4.8	3	21	8.8	Sunrise
spt	spt	25	useable	58.5	31	7.5	22	7	6	4.1	2	30.5	15.5	Clovis?
percussion flute	spt	3	useable	58	28	6	17.5	6	5.5	4.7	1	23	12.1	Unknown fluted
sps	spt	7	reuse potential with repaired tip	55*	25	6	20.5	6	6	4.2	2	23	7.3	Plainview
pressure	spt then abrupt retouch	25	useable	56	28.5	5.5	22.5*	5.5	5.5	5.2	3	24.5	11.6	Sunrise
spt	spt	8	useable	57	29	5	24*	5	4	5.8	3	24	8.2+	Sunrise
spt with retouch	retouch	7	reuse potential with repaired tip	56*	22	5.5	20*	5	5	4	1	40.5	8.9+	Midland
sps	spt	39	exhausted	45	24	6	21	6	6	4	3.5	27	7.9	Unfluted Folsom?
press	press	38	useable	40	22	5	19	5	5	4.4	1	12	5.5	Sunrise-reworked?
perc?	perc?	7	useable	45	29	7	25	7	7	4.1	1	23	10.2	Clovis?
spt	sps	25	useable	42	26	6	22	6	6	4.3	2	13	7.3	Sunrise
spt	spt	25	useable	45	25	5	24	4.5	5	5	5	8	6.9	Sunrise
press	press	25	useable	45	25	5	21.5	5	5	5	2	18	6.6	Sunrise
percussion flute	spt	25	useable	50.5	29	5	25	5	5	5.8	5	12	8.8	Sunrise
sps	sps	7	reuse potential with repaired tip	35.5	24	5	22	5.5	5.5	4.8	2	11	5.9+	Sunrise
spt	spt	3	useable	43	22	5	18.5	5	5	4.4	1	20	6.4	Sunrise- reworked?
sps	pressure	25	useable	51	22	5	16	4.5	4.5	4.4	2	26	7.2	Plainview?
sps	perc?	7	useable	62	26	6	20	6	6	4.3	3	34	12.8	Plainview
spt	spt	3	useable	41	20.5	5.5	17	5.5	5	3.7	0.5	22	5.4	Plainview-reworked
press	spt	25	useable with minor tip work	47*	27	5.5	25	5.5	5.5	4.6	3	15	8.3	Sunrise
perc scar from side	sps with retouch	25	reuse potential with redone tip	55	26	5.5	24	5.5	5.5	4.7	5	17	8.8+	Sunrise
sps	sps	7	useable	78	28	7	19	6	6	4	1.5	29	19.3	Plainview
sps	spt	7	reuse potential with repaired tip	62.5*	27	6	19	6	5	4.5	2	33.5	12.5	Plainview?
spt	spt	32	unfinished	49	26	4.5	23.5	4.5	4.5	5.2	3	14	7.7	Sunrise
missing	missing	25?	exhausted	54	n/a	5	n/a	n/a	5	n/a	n/a	n/a	7.6+	Sunrise
sps	spt	25	useable	48	27	5	24.5	5	5	4.9	1.5	16	6.7	Sunrise
spt	sps	25	reuse potential with repaired tip	61*	26	6	22	6	6	4.3	3	18	11.6	Sunrise
spt	spt	25	useable	46	27	4.5	21	4.5	4.5	6	5	22	7.6	Sunrise
spt	spt	25	useable	57	26	5	24	5	5	5.2	2.5	22	7.9	Sunrise
spt	spt	34	reuse potential with redone tip	43*	24	5.5	22	5	5	4.4	3	15	6.2	Sunrise
?	?	25	exhausted	53*	28*	5	26*	5	5	5.6	7*	16*	7.5+	Sunrise
sps	sps	3	useable	43	24	5	21	5	5	4.8	1	20	6.7	Sunrise
spt	spt	25	reuse potential with redone tip	53*	26	5	23	5	5	5.2	3	13	7.9+	Sunrise
press	press	32	useable	61	30	7	*21	7	7	4.3	1	30	16.2*	Plainview
sps	sps	25?	useable	61	30	5	20	5	5	6	2	28	12.3	Sunrise
percussion flute	sps	25	reuse potential with redone base	62	29	5.5	25*	5.5	5	5.3	5	15	12.3+	Sunrise
sps	sps	25	useable	46	26.5	6	24	5.5	6	4.4	2.5	22	8.6	Sunrise
sps	pressure	3	potential to be finished	56	25	5	19.5	5	5	5	1	31	10.1	Plainview?
none	spt	25	use potential before break	89	26.5	5	23.5	5	5	5.3	4	37.5	15.3	Goshen
sps	sps	24	useable	72	29	8	22	7.5	6.5	3.6	1.5	28.5	19.7	Plainview
spt-sps	sps	25	useable	45	22.5	5.5	19	5.5	5	4.1	2	16	6.8	Sunrise
spt	spt	9	reuse potential with redone base	64	24	5	24.5*	n/a	5	4.8	6	0	6.5*	Frontier Complex?
sps	spr	25	useable	49	25.5	6	21.5	5	5	4.3	5	15	7.6	Sunrise
spt	sps	24	reuse potential with repaired tip	64*	25.5	5	20	4.5	4.5	5.1	2	26	10.1*	Sunrise

Technology codes
spt= selective pressure thinning
sps= selective pressure shaping
ppt=percussion and pressure thinning
pps=percussion and pressure shaping
ppr-percussion pressure retouch
spr= selective pressure retouch

base forms in Plainview book p.268 Fig 11.9

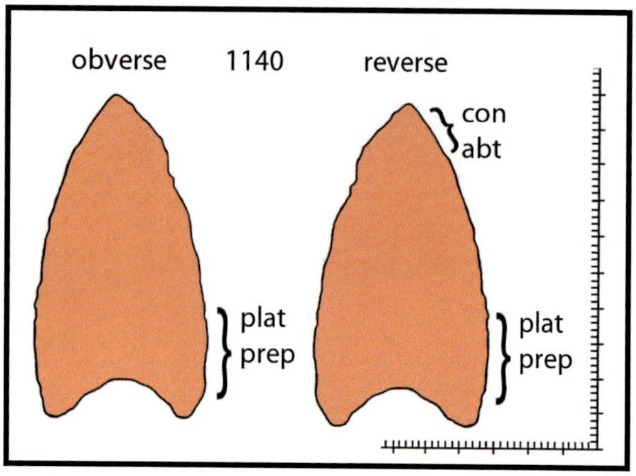
obverse 1140 reverse
con
abt
plat
prep
plat
prep

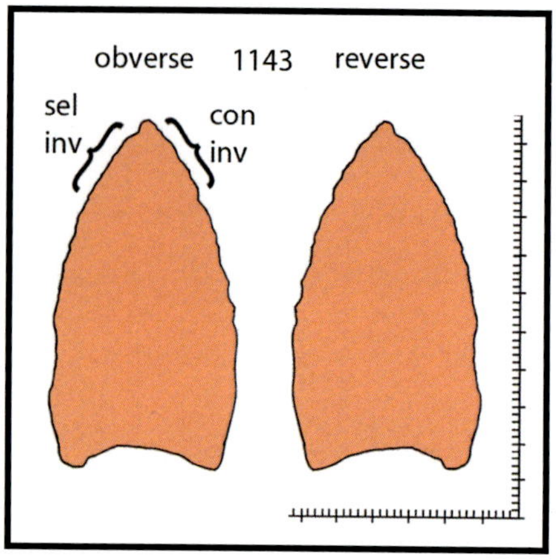
obverse 1143 reverse
sel
inv
con
inv

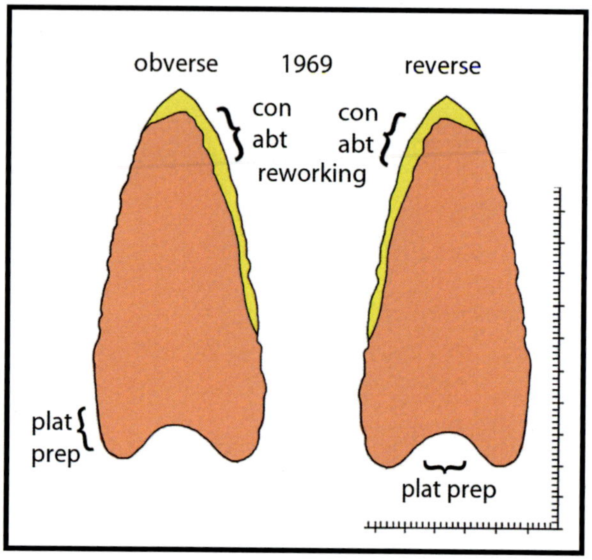
obverse 1969 reverse
con
abt
reworking
con
abt
plat
prep
plat prep

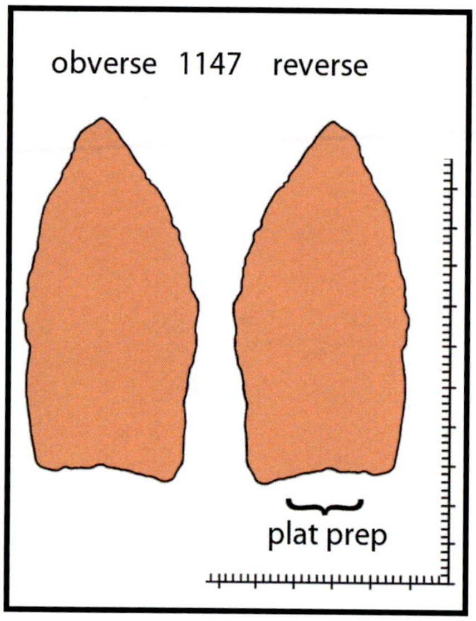
obverse 1147 reverse
plat prep

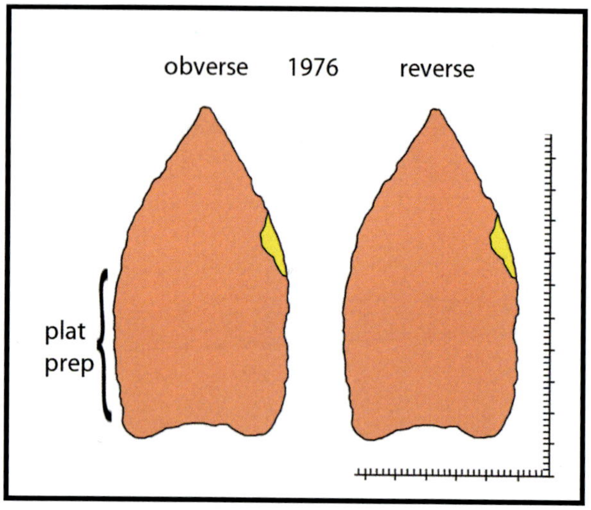
obverse 1976 reverse
plat
prep

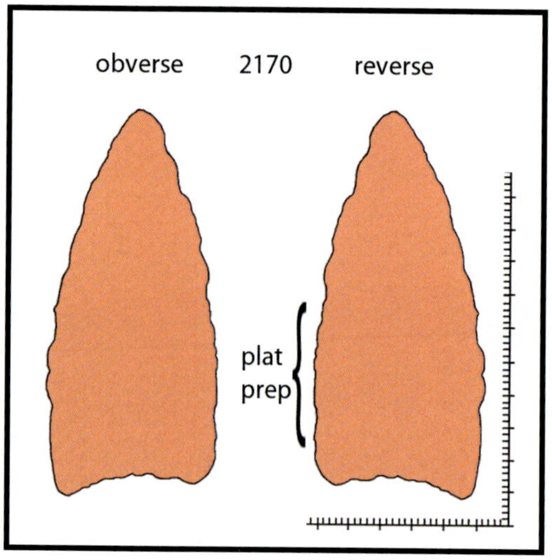
obverse 2170 reverse
plat
prep

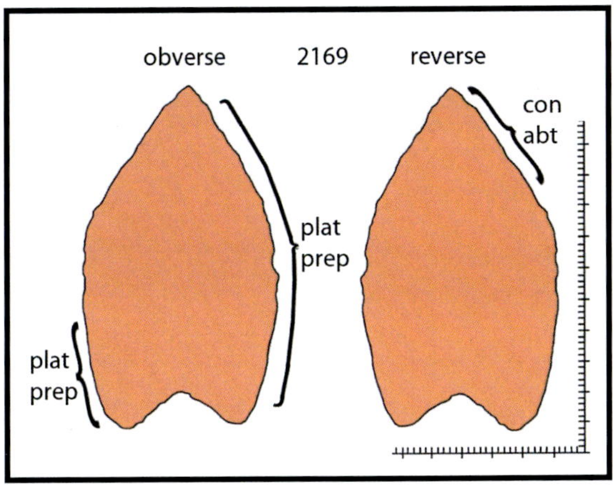
obverse
2169
reverse
con
abt
plat
prep
plat
prep

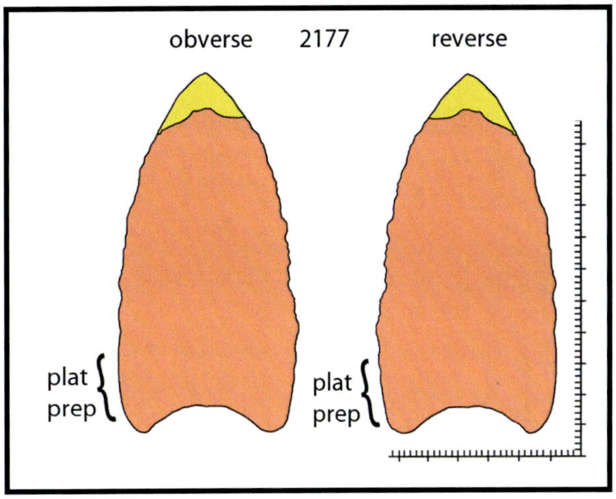
obverse
2177
reverse
plat
prep
plat
prep

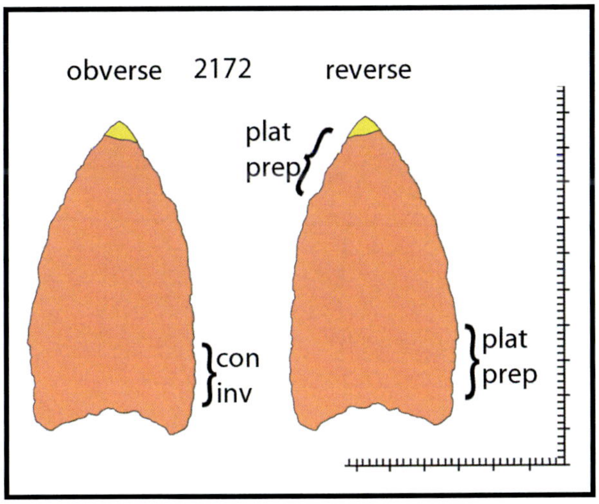
obverse
2172
reverse
plat
prep
con
inv
plat
prep

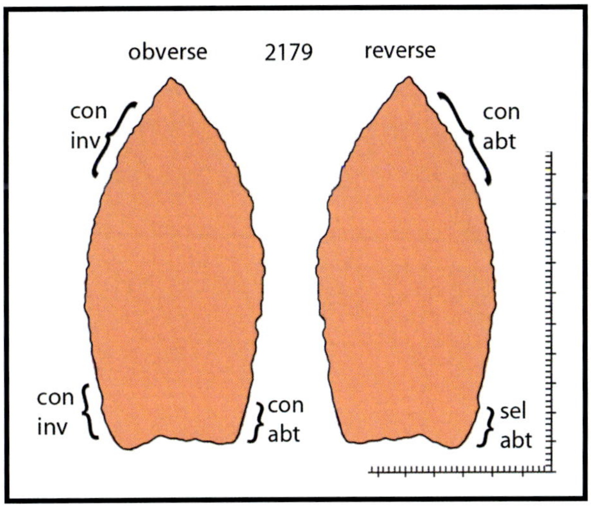
obverse
2179
reverse
con
inv
con
abt
con
inv
con
abt
sel
abt

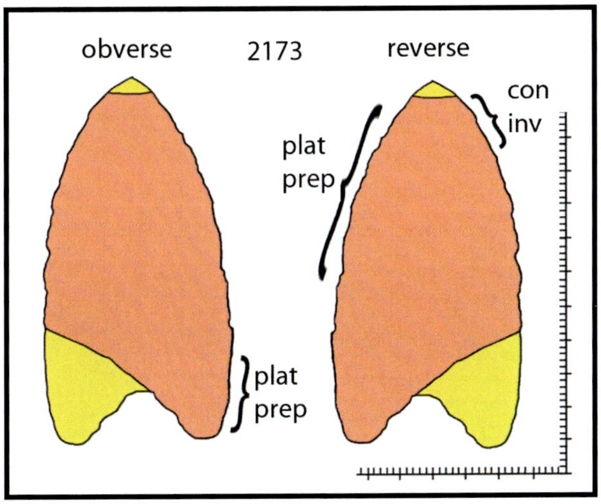
obverse
2173
reverse
con
inv
plat
prep
plat
prep

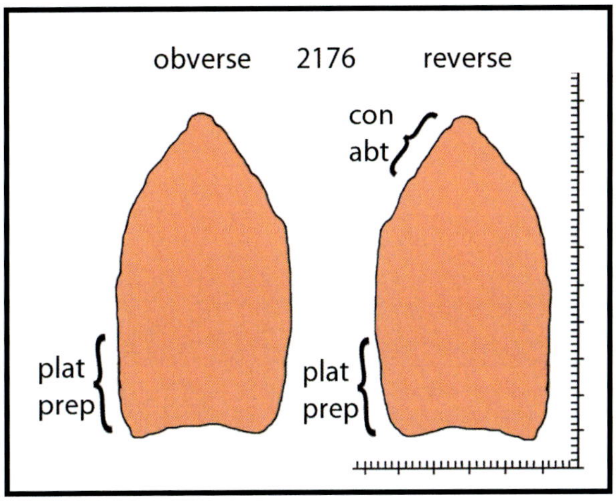
obverse
2176
reverse
con
abt
plat
prep
plat
prep

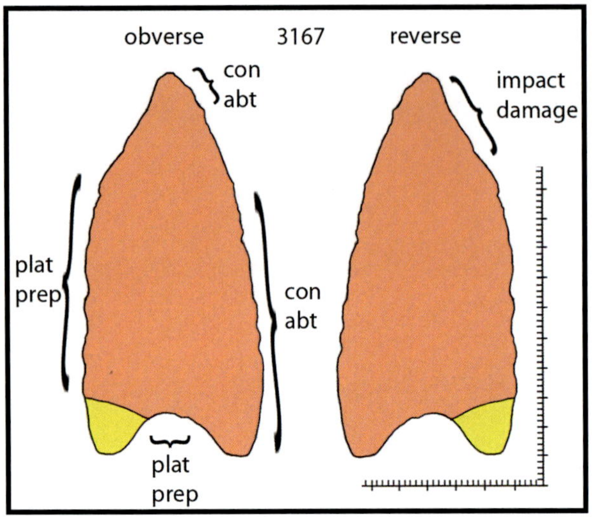
obverse
3167
reverse
con
abt
impact
damage
plat
prep
con
abt
plat
prep

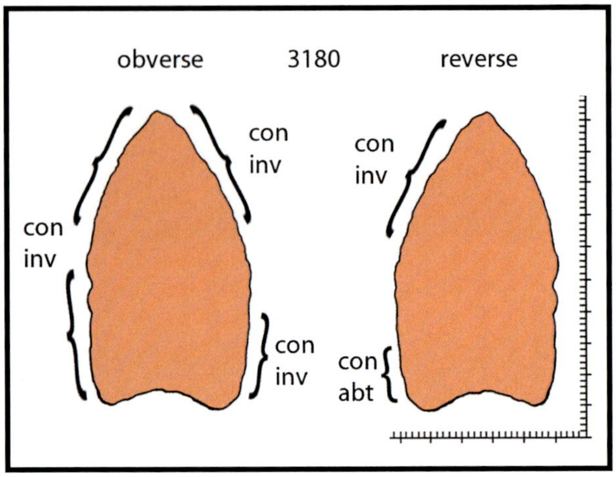
obverse
3180
reverse
con
inv
con
inv
con
inv
con
inv
con
abt

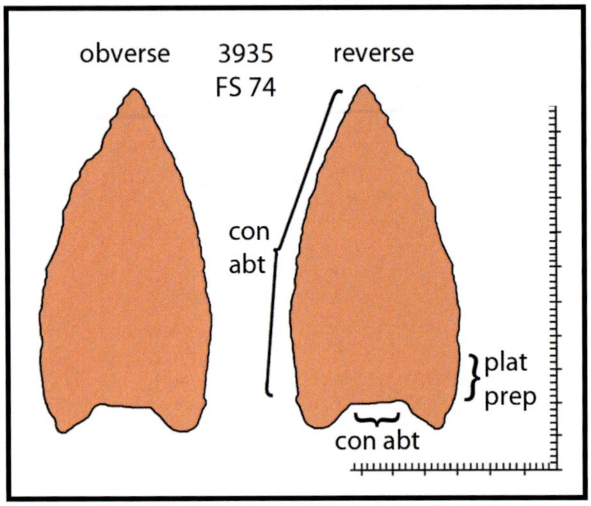
obverse
3935
FS 74
reverse
con
abt
plat
prep
con abt

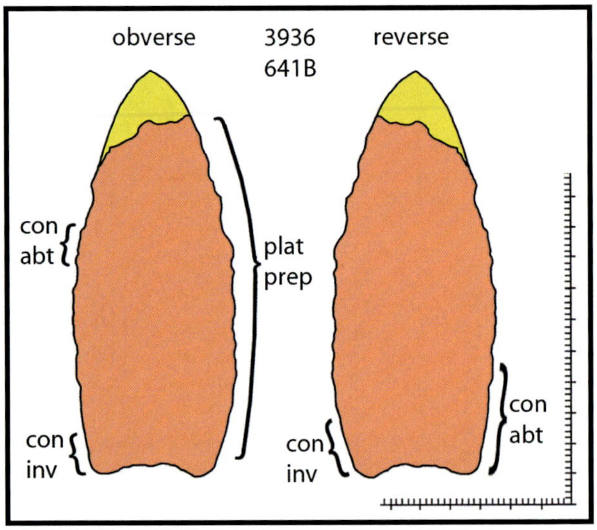
obverse
3936
641B
reverse
con
abt
plat
prep
con
inv
con
inv
con
abt

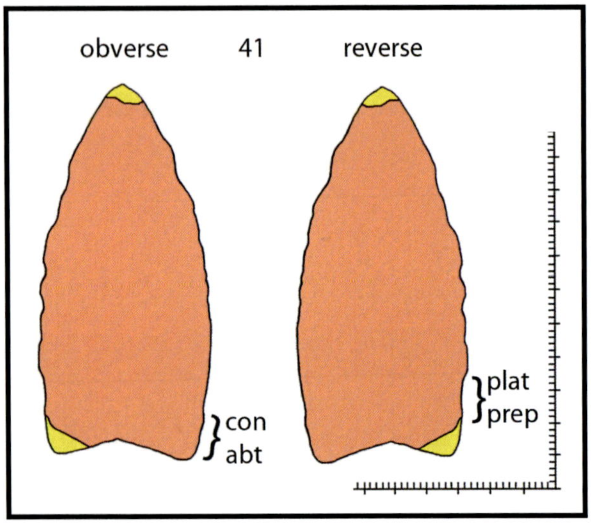
obverse
41
reverse
con
abt
plat
prep

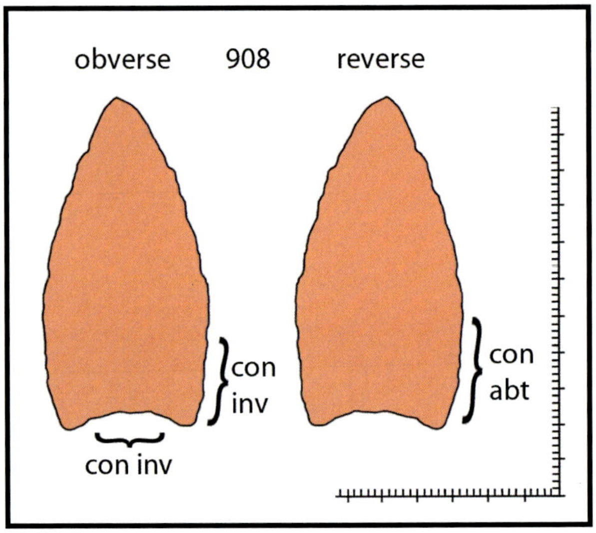
obverse
908
reverse
con
inv
con inv
con
abt

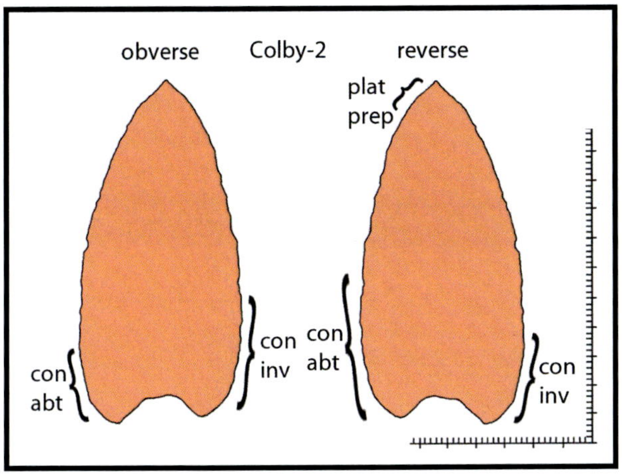
obverse
Colby-2
reverse
plat
prep
con
abt
con
inv
con
abt
con
inv

Chapter 8

Appendix 1

Examination of Group Membership

The membership of groups that composed Group 3 and Group 4 were investigated to determine if the Sunrise points would separate from the other paleo-points if more groups were used. Figure A8-1.1 shows the dendrogram and indicates the labels that are used to show the membership of each group. The membership of Group 3 and Group 4 is shown in Table A8-1.1.

Figure A8-1.1 Dendrogram showing labels of the groups examined.

Table A8-1.1 Membership of Group 3 and Group 4

(Taken from Chapter 7)

Group 3	Group 4
48PL330-1114.jpg	48PL330-1115.jpg
48PL330-1116.jpg	48PL330-1118.jpg
48PL330-1119.jpg	48PL330-1126.jpg
48PL330-1123.jpg	48PL330-1134.jpg
48PL330-1131.jpg	48PL330-1135.jpg
48PL330-1132.jpg	48PL330-1136.jpg
48PL330-1140.jpg	48PL330-1137.jpg
48PL330-1941.jpg	48PL330-1138.jpg
48PL330-1970.jpg	48PL330-1143.jpg
48PL330-2169.jpg	48PL330-1146.jpg
48PL330-2173.jpg	48PL330-1147.jpg
48PL330-2178.jpg	48PL330-1158.jpg
48PL330-2179.jpg	48PL330-1963.jpg
48PL330-3167.jpg	48PL330-1968.jpg
48PL330-3388.jpg	48PL330-1969.jpg
48PL330-3936-FS6418.jpg	48PL330-1972.jpg
48PL330-823.jpg	48PL330-1976.jpg

Cby-1.jpg	48PL330-2000.jpg
Cby-2.jpg	48PL330-2170.jpg
Cby-3.jpg	48PL330-2172.jpg
Cby-4.jpg	48PL330-2176.jpg
Csp.jpg	48PL330-2177.jpg
Dent.jpg	48PL330-2183.jpg
Jfc-1.jpg	48PL330-3180.jpg
MS-1.jpg	48PL330-3404.jpg
MS-2.jpg	48PL330-391-FS-908.jpg
MS-5.jpg	48PL330-3935FS74.jpg
Plv-TMM725-2.jpg	48PL330-41.jpg
UnkC-2.jpg	48PL330-45.jpg
WS.jpg	Lap-1.jpg
Wyc-2.jpg	MI-1.jpg
Wyc-3.jpg	MI-2.jpg
Wyo.jpg	MI-4.jpg
	MI-5.jpg
	MI-8.jpg
	Mid-1.jpg
	MS-3.jpg
	Pin.jpg
	Plv-1.jpg
	Plv-CS-1.jpg

Group 3 is composed of Group 3a and 3b. The membership of those groups is shown in Table A8-1.2. Group 3b contains all the Sunrise points from Group 3 but still contains some of the other paleo-points.

Table A8-1.2

Group 3a	Group 3b
Cby-3.jpg	48PL330-1114.jpg
Cby-4.jpg	48PL330-1116.jpg
Dent.jpg	48PL330-1119.jpg
Jfc-1.jpg	48PL330-1123.jpg
MS-1.jpg	48PL330-1131.jpg
Wyc-2.jpg	48PL330-1132.jpg
Wyc-3.jpg	48PL330-1140.jpg
Wyo.jpg	48PL330-1941.jpg
	48PL330-1970.jpg
	48PL330-2169.jpg
	48PL330-2173.jpg
	48PL330-2178.jpg
	48PL330-2179.jpg
	48PL330-3167.jpg
	48PL330-3388.jpg
	48PL330-3936-FS641B.jpg
	48PL330-823.jpg
	Cby-1.jpg
	Cby-2.jpg
	Csp.jpg
	MS-2.jpg
	MS-5.jpg
	Plv-TMM725-2.jpg
	UnkC-2.jpg
	WS.jpg

Group 3b is composed of Group 3b1 and Group 3b2. The membership of these groups is given in Table A8-1.3. Clearly, more of the other paleo-points are separated from the Sunrise points, however, two of the Sunrise points are grouped with those paleo-points. These results indicate that many of the Group 3 Sunrise points are very similar but that five other points, not believed to be Sunrise points, are also similar.

Table A8-1.3 Membership of Group 3b1 and Group 3b2

Group 3b1	Group 3b2
48PL330-1119.jpg	48PL330-1114.jpg
48PL330-1941.jpg	'48PL330-1116.jpg
48PL330-1970.jpg	48PL330-1123.jpg
48PL330-2178.jpg	48PL330-1131.jpg
48PL330-3936-FS641B.jpg	48PL330-1132.jpg
Cby-1.jpg	48PL330-1140.jpg
Csp.jpg	48PL330-2169.jpg
MS-5.jpg	48PL330-2173.jpg
Plv-TMM725-2.jpg	48PL330-2179.jpg
UnkC-2.jpg	48PL330-3167.jpg
	48PL330-3388.jpg
	48PL330-823.jpg
	Cby-2.jpg
	MS-2.jpg
	WS.jpg

Group 4 is composed of Group 4a and Group 4b. The membership of those groups is given in Table A8-1.4. Sunrise points from Group 4 are contained in both Group 4a and Group 4b as well as other paleo-points that were not Sunrise points. Group 4 is composed of points that do separate Sunrise from other paleo-points in the group. Rather, it appears that Group 4 is composed of two groups that each contain Sunrise and paleo-points.

Table A8-1.4 Membership of Group 4a and Group 4b

Group 4a	Group 4b
48PL330-1137.jpg	48PL330-1115.jpg
48PL330-1138.jpg	48PL330-1118.jpg
48PL330-1968.jpg	48PL330-1126.jpg
48PL330-1969.jpg	48PL330-1134.jpg
48PL330-1976.jpg	48PL330-1135.jpg
48PL330-2176.jpg	48PL330-1136.jpg
48PL330-2177.jpg	48PL330-1143.jpg
48PL330-2183.jpg	48PL330-1146.jpg
48PL330-391-FS-908.jpg	48PL330-1147.jpg
48PL330-3935FS74.jpg	48PL330-1158.jpg
48PL330-45.jpg	48PL330-1963.jpg
Mid-1.jpg	48PL330-1972.jpg
	48PL330-2000.jpg
	48PL330-2170.jpg
	48PL330-2172.jpg
	48PL330-3180.jpg
	48PL330-3404.jpg
	48PL330-41.jpg
	Lap-1.jpg
	MI-1.jpg
	MI-2.jpg
	MI-4.jpg
	MI-5.jpg
	MI-8.jpg
	MS-3.jpg
	Pin.jpg
	Plv-1.jpg
	Plv-CS-1.jpg

This examination indicates that the number of other paleo-points associated with Sunrise points in Group 3 could be reduced by breaking the point in the dataset into more groups, that is not true for Group 4.

Chapter 9

Appendix 1

Agate Basin Projectile Points.

Frontier Projectile Points.

Hell Gap Projectile Points.

References

Adams, Larry
2018 A Lapidary's View on John Voight Chert. Report to George Frison and George Zeimens.

Agenbroad, Larry D.
1978 *The Hudson-Meng Site: An Alberta Bison Kill in the Nebraska High Plains*. University Press of America. Washington, D.C.

Agogino, George A.
2009 The First Two Years of Investigation, 1959-1961: The Discovery and How the Later Investigation Leadership Developed. In *Hell Gap: A Stratified Paleoindian Campsite at the Edge of the Rockies*. Edited by Mary Lou Larson, Marcel Kornfeld and George C. Frison. The University of Utah Press, Salt Lake City. Pp 316-320.

Ahler, Stanley A.
1979 Functional Analysis of Nonobsidian Chipped Stone Artifacts: Terms, Variables and Qualifications. In *Lithic Use-Wear Analysis*. Edited by Brian Hayden, pp 301-317. Academic Press, New York.

1965 The Sister's Hill Site: A Hell Gap Site in North-Central Wyoming. *American Antiquity* Vol. 10: 190-195.

Ball, Sydney H.
1906 The Hartville Iron-Ore Range, Wyoming. In *Contributions to Economic Geology*. Department of the Interior, US Geological Survey, Government Printing Office. Washington, DC. pp 190-205.

Barbour, E.H. and Bertrand Schultz
1932 *The Scottsbluff Bison Quarry and its Artifacts*. Bulletin 34. Nebraska State Museum, Lincoln.

Bass, Mrs. Charles
1962 Chapter One - Red Men. *The Torrington Telegram*, Monday December 10.

Berner, John F.
2007 Announcing an Important Paleo Discovery: The Dr. F.L. Riser Paleo Cache. *Central States Archeological Journal*. Vol. 54, NO. 4, pp 98-101.

Boles, Steven J.
2021 Wyoming Road Trip: Remembering George C. Frison. *Central States Archeological Journal*. 68(2), pp 98-101.

Boles, Steven J., Marvin Kay and George C. Frison
2008 Relating the Dr. Riser-Healy Cache to the Colby Complex. Paper Presented at the 66th Annual Plains Conference, Laramie, Wyoming.

Bradley, Bruce A.
2009 Bifacial Technology and Paleoindian Projectile Points. In *A Stratified Paleoindian Campsite at the Edge of the Rockies*. Edited by Mary Lou Larson, Marcel Kornfeld and George C. Frison, pp 259-263. University of Utah Press, Salt Lake City.

Bradley, Bruce A., Michael B. Collins and Andrew Hemmings
2010 *Clovis Technology*. International Monographs in Prehistory. Archeological Series 17 pp 10, Ann Arbor, Michigan.

Bradley, Bruce A. and George C. Frison
1996 Flaked Stone and Worked-Bone Artifacts from the Mill Iron Site. In *The Mill Iron Site*. Edited by George C. Frison, pp43-69. University of New Mexico Press, Albuquerque.

Bronson, Edgar Beecher
1962 *Reminiscences of a Ranchman*. University of Nebraska Press, Lincoln.

Brooks, A.S., J.E. Yellen, R. Potts, A.K. Behrensmeyer, A.L. Deino, D.E. Leslie, S.H Ambrose, J.R Ferguson, F.d'Errico, A.M. Ziplin, S. Whittaker, J. Post, E.G. Veatch, K. Foecke and J.B. Clark.
2018 Long Distance Stone Transport and Pigment Use in the Earliest Middle Stone Age. *Science*, 360, pp. 90-94.

Buchanan, Briggs and Mark Collard
2010 A Geometric Morphometrics-Based Assessment of Blade Shape Differences Among Paleoindian Projectile Point Types From Western North America. *Journal of Archeological Science*, 37.

Buchanan, Briggs, Mark Collard and Michael J. O'Brien
2019 Geometric Morphometric Analyses Support Incorporating the Goshen Point Type into Plainview. *History Faculty Publications*, 27.

Callahan, Errett
1979 The Basics of Biface Knapping in the Eastern Fluted Point Tradition: A Manual for Flintknappers and Lithic Analysts. *Archeology of Eastern North America 7*(1).

Chance, H.M.
1901 The Iron Mines of Hartville, Wyoming. *Transactions of the American Institute of Mining Engineers*, Vol. XXX. Pp 987-1003. New York City.

Christiansen, Robert C.
1990 Neutron Activation Characterization of Knife River Flint from the Primary Source Area in West Central North Dakota. Paper presented at the 48th Plains Anthropological Conference, Oklahoma City.

Collins, Michael B.
1999 *Clovis Blade Technology*. Pp 7-10. University of Texas Press, Austin.

Davis, Leslie B.
1973 Schmitt (24BW559) A Middle Period Mine at the Missouri River Headwaters, Montana. Paper presented at the Thirty-First Plains Conference at Colombia, Missouri, 1973.

Dayet, Laure
2003 Defining Iron-Bearing Rock Sources in Compositional Data Statistical Analysis for Provenance Studies: A Case Study From South Africa. Journal Archeological Science: Reports Vol. 49, June.

Eddy, J.A.
1977 Mystery of the Medicine Wheels. *National Geographic* 15(1):140-146.

Emmons, S.F. and E.C. Eckel
1907 The Hartville Iron-Ore Range, Wyoming. *Contributions to Economic Geology*. Pp 190-205. Department of the Interior, United States Geological Survey, Washington, D.C.

Evilsizer, Laura
2016 Knife River Flint Distribution and Identification in Montana. Masters Thesis, University on Montana, Missoula.

Figgins, Jesse D.
1927 The Antiquity of Man in America. *Natural History*, 27 (3).

Frey, Eugene
1947 Hartville Iron District, Platte County, Wyoming: U.S. Bureau of Mines, Report of Investigations 4086, p 3.

Frison, George C.
1974 *The Casper Site: A Hell Gap Bison Kill on the High Plains*. Academic Press, New York.

1984 The Carter/Kerr-McGee Paleoindian Site: Cultural Resource Management and Archeological Research. *American Antiquity*, 49. (2).

1991 *Prehistoric Hunters of the High Plains*, second edition, pp 291 Academic Press, San Diego, California.

1996 *The Mill Iron Site*. University of New Mexico Press. Albuquerque.

2004 *Survival by Hunting*. Pp 219. University of California Press, Berkeley.

Frison, George C., Vance Haynes, Jr. and Mary Lou Larson
1996 Discussions and Conclusions. In *The Mill Iron Site*. Edited by George C. Frison, pp 205-216. University of New Mexico Press, Albuquerque.

Frison, George C. and Bruce A. Bradley
1980 *Folsom Tools and Technology at the Hanson Site, Wyoming*. University of New Mexico Press, Albuquerque.

1999 *The Fenn Cache: Clovis Weapons and Tools*. One Horse Land and Cattle Company, Santa Fe, New Mexico.

Frison, George C. and Dennis Stanford (Editors)
1982 *The Agate Basin Site: A Record of Paleoindian Occupation on the Northwestern High Plains*. Academic Press, New York.

Frison, George C. and Lawrence C. Todd
1987 *The Horner Site, The Type Site of the Cody Cultural Complex*. Academic Press, Orlando.

1986 *The Colby Mammoth Site: Taphonomy and Archeology of a Clovis Kill in Northern Wyoming*. University of New Mexico Press, Albuquerque.

Frison, George C., George M. Zeimens, Spencer Pelton, Danny Walker, Dennis Stanford and Marcel Kornfeld.
2018 Further Insights into Paleoindian Use of the Powars II Red Ochre Quarry, (48PL330) Wyoming. *American Antiquity*, Vol. 83, No. 3. pp 485-504.

Gilder, Robert F.
1908 Indian Sites Near Frederick, Wyoming. *Records of the Past*. Vol. 7 Part 4, Omaha, Nebraska.

1909 The "Spanish Diggings, Wyoming". *Records of the Past*. Vol. 8, Part 1, Omaha, Nebraska.

Gourhan, Leroi A.
1968 *The Art of Prehistoric Man in Western Europe*. Thames and Hudson, London.

Harris, W.D., J.F. McLaughlin and R.W. Jones
2005 Geologic Map of the Guernsey Quadrangle, Platte and Goshen Counties, Wyoming. Wyoming State Geological Survey Map Series 68, scale 1:24,000.

Haury, Emil W., E.B. Sayles and W.H. Wasley
1959 The Lehner Mammoth Site, Southeastern Arizona. *American Antiquity* 25(1):2-30.

Hausel, W. Dan
1989 The Geology of Wyoming's Precious Metal Lodes and Placer Deposits. *The Geological Survey of Wyoming*, Bulletin 68: 67,72.

Hausel, W.D. and W.M. Sutherland
2000 *Gemstones and Other Unique Minerals and Rocks of Wyoming - A Field Guide for Collectors*: Wyoming State Geological Survey Bulletin 71.

Hauser, Neil
2021 Using "Knowledge of the Crowd" to Inform Subjective and Objective Projectile Point Clustering. *Southwestern Lore*, Vol. 87, 3.

Haynes, C. Vance Jr.
1965 The Hell Gap Site, Wyoming: Geological Sketch. *Wyoming Archeologist* 8(2): 35.

1967 Carbon 14 Dates and Early Man in the New World. In *Pleistocene Extinctions: the Search for a Cause*. Edited by Paul Martin and H.E. Wright, Jr. Pp 267-286. York University Press.

2016 Personal Communication

Hodgskiss, Tammy
2013 Ochre Use in the Middle Stone Age at Sibudu, South Africa: Grinding, Rubbing, Scoring and Engraving. *Journal of African Archeology*. Vol. 11, No. 1 pp 75-95.

Hofman, Jack L., and Russell W. Graham
1998 The Paleo-Indian Cultures of the Great Plains. In *Archeology on the Great Plains*. edited by W. Raymond Wood. University of Kansas Press, Lawrence.

Holen, Kathleen and Steve R. Holen
2009 A Beveled Bone Rod from the Cody Component of the Lindenmeir Site. *Current Research in the Pleistocene*, 26: 65-67.

Holliday, Vance T., Eileen Johnson and Ruthann Knudson, editors
2017 *Plainview: An Enigmatic Artifact Style of the Great Plains*. The University of Utah Press. Salt Lake City.

Holmes, W.H.
1919 *Handbook of Aboriginal American Antiquities*. Pp 210-213 Bureau of American Ethnology Bulletin No. 60. Washington, DC.

Hovers, E., S. Ilami, D.Bar-Yosef and B. Vandermeesch
2003 An Early Case For Color Symbolism. *Current Anthropology*, Vol. 44 , pp 491-522.

Huckell, Bruse B.
2007 Clovis Lithic Technology: A View from the Upper San Pedro Valley. In *Murray Springs: A Clovis Site with Multiple Activity Areas in the San Pedro Valley, Arizona*. Edited by C. Vance Haynes and Bruce Huckell, pp 170-213. University of Arizona Press, Tucson.

Justice, Noel D.
2002 *Stone Age Spear and Arrow Points of the Southwestern United States*. Indiana University Press, Bloomington and Indianapolis, Indiana.

Irwin-William, C., G.H. Irwin, G. Agogino and C.V. Haynes
1973 Hell Gap: Paleoindian Occupation on the High Plains. *Plains Anthropologist*, 83(3): pp 485-504.

Kay, Marvin
1996 Microwear Analysis of Some Clovis Experimental Chipped Stone Tools. In *Stone Tools: Theoretical Insights into Human Prehistory*. Edited by G.H. Odell. 315-344, Plenum Press, New York.

Knudson, Ruthann
2017 The Plainview Assemblage in Context. In *An Enigmatic Artifact Style of the Great Plains*. Edited by Vance T. Holliday, Eileen Johnson and Ruthann Knudson. The University of Utah Press, Salt Lake City pp. 29-78.

2008 Use-Wear Analysis for the Jim Pitts Site: 39CU1142, South Dakota. Report to the South Dakota Historical Society, Rapid City.

Knight, Wilber C.
1901 Iron and Copper Economics. *The Wyoming Industrial Journal*. Vol. 2, No. 10. Cheyenne, Wyoming, pp 64.

Kornfeld, Marcel, George C. Frison and Mary Lou Larson
2010 *Prehistoric Hunter-Gatherers on the High Plains and Rockies*, pp 451, Third Edition, Left Coast Press, Walnut Creek, California.

Koski, Randolph
2012 Supreme and Gangue Characteristics in Volcanogenic Massive Occurrence Model. U.S. Geological Survey, Scientific Investigation Report 2010-5070-C, p 186.

Kristensen, Todd, Emily Moffat, John M. Duke, Andrew J. Locock, Cody Sharphead and John W. Ives
2018 Identifying Knife River Flint in Alberta: A Silicified Lignite Toolstone from North Dakota. *The Swing of Things: Contributions to Archeological Research in Alberta*. Archeological Survey of Alberta, Occasional Paper No. 38.

Lahren, Larry A., and Robson Bonnichsen
1974 Bone Foreshafts From a Clovis Burial in Southwestern Montana. *Science*, Vol. 186, pp 148-150.

Larson, Mary Lou, Marcel Kornfeld and George C. Frison
2009 *Hell Gap, A Stratified Paleoindian Campsite at the Edge of the Rockies*. University of Utah Press, Salt Lake City.

MacCannon, R.W.
2003 *Sunrise: A Chronology of a Wyoming Mine*. Published by the Bessemer Historical Society, Pueblo, Colorado. pp 9.

Mackie, Madeline E., Todd Suravell, Matthew O'Brien, Robert Kelly, Spencer Pelson. C. Vance Haynes Jr., George C. Frison, Robert Yoke, Steve Teteak, Heather Rockwell and Shannon Mahan
2020 Confirming a Cultural Association at the La Prele Mammoth Site (48CO1401) Converse County, Wyoming. *American Antiquity*. Vol. 85, (3) 554-572.

Miller, James C.
2010 Lithic Resources. In *Prehistoric Hunters of the High Plains and Rockies*, 3rd edition. Left Coast Press, Walnut Creek, CA pp 596.

Millgate, M.L.
1965 The Haystack Range, Goshen and Platte Counties, Wyoming: Wyoming State Geological Survey, Preliminary Report.

Morrow, Juliet E.
2016 Evidence for Paleoindian Spirituality and Ritual Behavior: Large Thin Bifaces and Other Sacred Objects from Clovis and Other Late Pleistocene-Early Holocene Cultural Contexts. In *Research, Preservation, Communication: Honoring Thomas J. Green on His Retirement from the Arkansas Archeological Survey*. Edited by Mary Beth Trubitt. Pp 18-65. Research Series No. 67, Arkansas Archeological Survey, Fayetteville.

Nami, Hugo
2014 Observaciones para onocer secuencias de reduccion bifaciales paleoindias y puntas Fwll en el valle del Ilalo, Ecuador. In Poblacion de America del Sur: la contribucion de la technologia lithica. *Proceedings of the 16th World Congress International Union of Prehistoric and Protohistoric Sciences*. Archeoeditions, Lormont, FR.

Pelton, Spencer R., Lorena Becerra-Valdivia, Alexander Craib, Sarah Allaun, Chase Mahan, Charles Koenig, George Zeimens and George C. Frison
2022 In Situ Evidence for Paleoindian Hematite Quarrying at the Powars II Site (48PL330), Wyoming, USA. *Proceedings of the National Academy of Sciences*. Vol. 119, No. 20.

Reher, Charles A.
1991 Large Scale Lithic Quarries and Regional Transport Systems on the High Plains of Eastern Wyoming. In *Raw Material Economies Among Prehistoric Hunter-Gatherers*, Edited by A. Montet-White and Steve Holen, University of Kansas Publications in Anthropology No. 19, Lawrence. Pp 251-284

Rigaud, Jean-Philippe
1988 Art Treasures from the Ice Age Lascaux Cave. *National Geographic*, Vol. 174, No. 4.

Roebroeks, Wil, Mark J. Sier, Trine Kellberg Nielsen, Nimitri De Loecker, Josep Maria Par‘s, Charles E. Arps and Herman J. Mu‘cher.
2012 Use of Red Ochre by Early Neanderthals. *Proceedings of the National Academy of Sciences*. Vol. 109, No. 6.

Rohif, F. James
2004 tpsDig. *Digitized Landmarks and Outlines, Version 2.0*. Department of Ecology and Evolution, State University of New York at Stony Brook.

Root, M.J., J.D. Williams, M. Kay and L.K. Shifrif
1999 Folsom Ultrathin Biface and Radical Break Tools in the Knife River Flint Quarry Area. In *Folsom Lithic Technology: Exploration in Structure and Variation*. Edited by D.S. Amick. Archeological Series 12, pp 144-168 *International Monographs in Prehistory*. Ann Arbor, Michigan.

Roper, Donna C.
1991 A Comparison of Contexts of Red Ochre Use in Paleoindian and Upper Paleolithic Sites. *North American Archeologist*, pp 289-301 Vol. 12(4).

Rulenbeek, F.J.A.
2007 Hydrothermal Processes in the Archean from Imagining Spectoscopy. *International Institute for Geo-Information Science and Earth*. Pp 79, 92. Enchede, Netherlands.

Rupp, G.H.
1939 Block Caving at the Sunrise Iron Mine, Wyoming. *AIME Technical Publications*, No. 1067. P 16.

Saul, John M.
1969 Study of the Spanish Diggings, Aboriginal Flint Quarries of Southwest Wyoming. *National Geographic Research Reports 1964 Projects*, pp 183-199. Washington D.C.

Sims. P.K., W.C Day, G.L. Snyder and A.B Wilson, with a section on Geochronology of post-tectonic intrusive rocks by Zell E. Peterman
1997 Geologic Map of Precambrian Rocks Along Parts of the Hartville Uplift, Guernsey and Casebier Hill Quadrangles, Platte and Goshen Counties, Wyoming. *Miscellaneous Investigations Series, Map 1-2567 Pamphlet*. US Geological Survey, US Department of the Interior.

Smith, Heather
2010 Geometric Morphometrics and Clovis Points: Searching for Behavorial Patterns, Texas Beyond History. https://www.texasbeyondhistory.net/spotlights/geomet-morphomet/geomet-morphomet.html

Snow, E.R.
1895 The Hartville Iron Ore Deposits in Wyoming. In *The Engineering Mining Journal*, pp 320-321. Vol. 60. University of Michigan.

Snyder, George L.
1980 *Map of Precambrian and Adjacent Phanerozoic Rocks of the Hartville Uplift, Goshen, Niobrara and Platte Counties, Wyoming*. Open File Report 80-779. US Department of the Interior Geological Survey.

Stafford, Michael D., George C. Frison, Dennis Stanford, and George Zeimens.
2003 Digging for the Color of Life: Paleoindian Red Ochre Mining at the Powars II Site, Platte County, Wyoming. *Geoarchaeology* 18: pp 71-90.

Stanford, Dennis J.
1974 Preliminary Report on the Excavation of the Jones-Miller Hell Gap Site, Yuma County, Colorado. *Southwestern Lore*, 40(3-4):29-36.

Sutherland, W.M., Gregory R.W. Carnes, J.D. and B.N. Worman
2013 Rare Earth Elements in Wyoming: Wyoming State Geological Survey Report of Investigations 65.

Sutherland, W.M. and E.C. Cola
2016 Iron Resources in Wyoming: Wyoming State Geological Survey Report of Investigations 67.

2016 A Comprehensive Report on Rare Earth Elements in Wyoming: Wyoming State Geological Survey Report of Investigations, 71.

Tankersley, K.B., N.R. Shafer, M.D. Hess, J.S. Benz, F. Randolph Turner, M.D. Stafford, G.M. Zeimens and G.C. Frison.
1995 They Have a Rock That Bleeds: Sunrise Red Ochre and its Early Paleoindian Occurrence at the Hell Gap Site, Wyoming. *Plains Anthropologist*, 40: pp 185-194.

Tarlach, Gemma
2018 Pigment of Our Imagination: The story of Human Evolution Is Written in Ochre. *Discover, Science for the Curious*. April, Vol. 39. No. 3.

Taylor, D.C.
1969 The Wisall Excavations: An Exercise in Frustration. *Proceedings of the Montana Academy of Sciences*. -15029. 147.

Waldorf, D.C.
2016 Lithic Materials from Sunrise Mine, Southeastern Wyoming. Report Prepared for Zeimens and Frison, Principal Site Investigators.

Weed, L.W.
1927 *CF&I Industrial Bulletin*, April.

Whissen, Steven
1985 Class III Cultural Resource Investigations of the Sunrise Mine. Department of Environmental Quality Abandoned Mine Land Project No. 10.

Wilkins, Elwood S.
1967 Problems in the Relation of "Flint" Tools Geological Place of Origin. *Bulletin of the Archaeological Society of Delaware*. Nos. 5-6. p 38.

Wilmsen. E.N., and F.H.H. Roberts
1978 Lindenmeier, 1934-1974. Smithsonian Contributions to Anthropology No.24. Washington, DC.

Wilson, Michael C., Kathie L. Road and Kenneth J. Hardy (editors)
1981 *Megaliths to Medicine Wheels: Boulder Structures in Archeology: Proceedings of the Annual Chacmool Conference*. The Archeological Association of the University of Calgary, Alberta.

Wilson, W.H.
1976 Wyoming Iron Resources Summary Report, unpublished report, U.S. Bureau of Mines, Contract No. JO265005.

Wormington, H.M. and R.G. Forbis
1965 An Introduction to the Archeology of Alberta, Canada. *Proceedings of the Denver Museum of Natural History*, Vol. 5, No. 11.

Wray, Charles F.
1946 Varieties and Sources of Flint Found in New York State. *Pennsylvanian Archeologist*. pp 25-45. Vol. XVII Nos. 1-2.

Wreschner, Ernst E.
1980 Red Ochre and Human Evolution: A Case for Discussion. *Current Anthropology*, (21) 5.

Zarycka, S.E., T.A Surovell, M.E. Mackie, S.R.Pelton, R.L. Kelly and P. Goldberg.
2019 Long-distance Travel Transport of Red Ochre by Clovis Foragers. *Journal of Archeological Science*, Reports 25, pp 519-529.

Zeimens. George M.
n.d. A Preliminary Study of the Archeological History and Prehistory of Cottonwood Draw in Northwestern Goshen County, Wyoming.

2002 A Story of Three Caves or Evidence for Early Archaic Adaptations in Eastern Wyoming. Prepared for the San Diego Foundation. On file at the Sunrise Historic and Prehistoric Preservation Society Headquarters, Sunrise, Wyoming.

2001 Preliminary Investigations of the Jewett Mammoth Site. In *Proceedings of the International Conference on Mammoth Site Studies*. Publications In Anthropology 22, University of Kansas, Lawrence.

1999 A Class III Cultural Resource Study for Expansion of the Union Pacific Railroad Between Lusk and Manville, Wyoming. Report on file at the State Historic Preservation Office.

Zeimens, George M., Geri McIver, Harry Earl, R. Clayton Housh, Dennis Eisenbarth, Garry Alkire and Dewey Baars.
1992 The Sommers Site: A Progress Report from the High Plains Chapter. *The Wyoming Archeologist*, 36(1-2): pp 11-25.

Zeimens, George M., Geri McIver, Dewey Baars, Alan and Terry Korell, Sandra Hansen, Harry Earl and R. Clayton Housh.
1995 Preliminary Report of Salvage Excavations at the Hell Gap Site, 48GO305, *The Wyoming Archaeologist*, Vol. 39 (1-2).

Zeimens, George M. and Dewey Baars
1996 Early Archaic Pit Features in East Central Wyoming. Paper presented at the 54th Plains Anthropological Conference, Iowa City.

Zeimens, George M. and Garry Alkire
2005 *Field Archeology: Basic Practices in the North Platte Valley*. Western Plains Historic Preservation Association, Lingle, Wyoming.

Newspapers
1899 *The Iron Gazette*, Workings of Prehistoric Mines. Vol. 1, No. 1, November 17.
Hartville Iron Deposits, Vol. 1., No. 2. November 24.

1902 *The Rawlins Republican*, October 29, page 4.

United States Federal Census 1899, 1900, 1920, Hartville, Wyoming.